LIVING ON THE REAL WORLD

HOW THINKING AND ACTING LIKE METEOROLOGISTS
WILL HELP SAVE THE PLANET   **WILLIAM H. HOOKE**

AMERICAN METEOROLOGICAL SOCIETY

Published by the American Meteorological Society
45 Beacon Street, Boston, Massachusetts 02108

For more AMS Books, see http://bookstore.ametsoc.org.

The mission of the American Meteorological Society is to advance the atmospheric and related sciences, technologies, applications, and services for the benefit of society. Founded in 1919, the AMS has a membership of more than 13,000 and represents the premier scientific and professional society serving the atmospheric and related sciences. Additional information regarding society activities and membership can be found at www.ametsoc.org.

Library of Congress Cataloging-in-Publication data is available and can be found online at www.ametsoc.org.

CONTENTS

To Jesus Christ and my wife Christine,
who came into my life as a package deal.

AUTHOR'S INTENT

... the design or intention of the author is neither available nor desirable as a standard for judging the success of a work of literary art. —*M. Beardsley and W.K. Wimsatt (1954)*

In the spring of 1961, I was in the second semester of my freshman year at Swarthmore College and taking philosophy 101. My professor was one Monroe Beardsley, a wonderfully thoughtful man, truly wise, and a gifted teacher. Among other things, Professor Beardsley instilled in us an appreciation for what he had labeled "the intentional fallacy." He reasoned, very persuasively, that an author was not necessarily the best judge of the meaning of his or her own work. He argued instead, as quoted above from his seminal paper, that the meaning of a work of art could be constructed only from the text itself and that the particulars of the author's biography or hopes and aspirations were completely superfluous.

You should therefore take this author's intent with a grain of salt. I may have partially, even grossly, missed the message of the book that I've written and that you hold here. On the other hand, with such fair notice, these few prefatory remarks should cause little additional harm. And every few months during the book's development, it proved useful to reflect on my purpose, which has evolved a bit from when I started.

1. First and foremost, the book calls attention to an extreme event now underway.

On the real world, extremes are really nothing new. Earth does much of its business through extremes. For example, consider continental *drift*. The word suggests an image of the continents moving at a pace of centimeters per year over hundreds of millions of years in response to underlying flows in the earth's mantle hundreds of miles beneath us. But from our perch atop the earth's crust, we experience this as a series of abrupt jerks and spasms: horrific earthquakes, their accompanying tsunamis, and active volcanism. In the same way, the atmosphere's circulation is gentle on average, but it is driven and abetted by violence: thunderstorms, tornadoes, hurricanes, blizzards, and ice storms. Most places on Earth don't experience continuous drizzle so much as brutal cycles of flood and drought: When it rains it pours.

Life on Earth, the plants and the animals in the sea and on the land, has adapted to these extremes. Indeed it relies upon them. Deserts bloom following a downpour. Many plant seeds germinate better after a wildfire. What's more, life also accomplishes its business through extremes. Ecologists tell us of energy flows and nutrient cycling, but look a little closer. Many prey species give birth all in the same short season, so that predators don't have time to staff up. Prey school and flock and swarm so as to pose other barriers to predation. One hint: Ecologists use the language of extremes. They speak of disease and pestilential *outbreaks*, of population *explosion* and *collapse*, of *die-off*.

But this extreme is different. It doesn't merely concern us, it *is* us. *We* are the extreme that matters. In a short period of time, the span of a few generations, we human beings have greatly added to our numbers, increased our resource use per capita, and stepped up the pace of science and social change. At first thought, this recent human success might not seem so consequential. But at the same time we have transformed the human problem of survival. For all of human experience up to now, the challenge of living on the real world has been twofold: meeting our needs for water, shelter, food, and other resources and protecting ourselves from natural hazards. The real world has been *resource* and *threat*. But our numbers and resource consumption are now so great that in the course of trampling around, though meaning no harm, we are transforming the landscape, reducing and altering habitat, decreasing biodiversity, and degrading the environment. Today and going forward, our problem is threefold. The real world is not only *resource* and *threat*, it is also *victim*.

We were barely smart enough, as individuals and as the human race, to solve our age-old twofold problem. Reaching this degree of success has taken hundreds of thousands of years and a lot of trial and error. Our day-to-day lives are in many cases and in many places shambolic.

The threefold problem is beyond us.

This extreme event, and our growing awareness of its implications, has given rise to a cottage industry of apocalyptic literature. Books and papers and newspaper articles and columns and television documentaries paint a grim picture of where we are and whither we are tending. The general story is often a variation or embellishment of the theme of bacteria on a Petri dish at the point at which the carrying capacity of the Petri dish is exceeded.

But we are not bacteria. And while we are not smart enough individually or even corporately to solve humanity's threefold problem in the way we are accustomed to think of solving problems, we have providentially found ourselves at a point in our scientific understanding and technological development where we have new tools for going forward. Which leads us to number 2.

2. This book is also a handbook.

It identifies four of these new tools. It outlines how together they might be used not as a *solution* to our threefold problem, which truly is not one we can solve, but as a *coping strategy*, a means of buying time—much as insulin injections don't cure diabetes but control it, allowing the patient to live a normal life. These four tools will be elaborated upon later, but let's enumerate them here:

1. *A basis of facts*: an unprecedented level of knowing and understanding the natural and social realities of the world we live on—a knowledge base that is accumulating all the time.
2. *Policy*: for transforming our understanding of those realities into frameworks for thinking and doing that simplify our living and enable us to make decisions and take action quickly and effectively.
3. *IT and social networking*: greatly multiplying the rate at which we are innovating, learning from experience, and sharing that learning with others.
4. *Responsibility and leadership*: at the individual and local level, devolving power and influence and more widely distributing leadership; turning mere conversation into action in communities across the world rather than just in a few national capitals or business hubs.

My intent is not to prescribe how these tools might be used so much as to awaken you to their possibilities. I want to stimulate you to learn more about these tools, including their strengths and limitations. I encourage you to master the use of these tools, singly and in combination, even to the point of being able to improve them. In this process, you will be doing your part to help seven billion people live more successfully on the real world.

3. Finally, much like a newspaper article, this book reports a breaking human story.

It's a story about you and me, and others of our generation, as we respond to the 21st-century's epic challenge. We're writing the narrative even as we actually live it out.

It's a big story that speaks to the true significance of the time we live in. It suggests that we are at one of history's sweet spots: a hinge point where everything we do matters, particularly how we work together, even when (or perhaps especially when) we don't know one another.

What's more, the "we" is all seven-going-on-nine billion of us, an entire generation: a group of not just my fellow atmospheric scientists, or a handful of people in the science policy world, or a narrow range of specialists in disaster risk management. Membership is not limited to a few political leaders or media moguls who shape the news we breathlessly follow or to people from any particular country or to those carrying some superior learning or expertise. It's all of us.

The narrative is not merely a news story but also an adventure story. If we succeed, our generation will be heroes. We will solve the 21st-century's most urgent problem: how to supply food, water, energy, and the other essentials of life for the world's people, while at the same time protecting the environment and building resilience to nature's hazards. We will *save* or, more accurately, *buy time* for civilization and humanity. We will make life worth living for generations to come; they will see and remember what we did. If we fail, the world may well decay, bit by bit, into a progressively darker, more despairing place. To succeed will require knowledge and understanding, but that's only the beginning. It will also require courage and nobility, honor and humility, as well as decisions and *action*—even selfless love—in every sphere of life: science and technology, politics and government at all levels, business large and small, education, public health, journalism, and more. As a result, part of the story is that you and I don't get to opt out, stay on the sidelines, or remain detached bystanders any more than our parents and grandparents (and great-grandparents) got to sit out World War II. Anyone alive then was necessarily involved. In much the same way today, every one of us is a major actor.

A successful outcome to this story will require our very best. And like all adventures, the story's theme transcends the natural and social worlds into the spiritual world as well.

Please dwell on this for a moment. This framing may seem a little over the top. *Twenty-first-century real-world living is an adventure story? With heroines and heroes, danger and risk, and momentous consequences hanging in the balance? A spiritual dimension? That doesn't square with what I see at my workplace or within my network of friends, or in my daily life, for that matter. That's certainly not what I've come to expect from books on resources or the environment or hazards. And, frankly, lumping the spiritual together with the scientific makes me uncomfortable.*

Point taken. Most of us see only a false urgency and an all-too-real futility in what we do. For a variety of reasons, we may be slow to realize where and in what ways we fit in or how much we individually matter. We might not see a clear path to how we make a difference. And some of us may have worked hard to rid our lives of any spiritual dimension. The overwhelming response we give when asked about our lives and work is usually some tired variant of *same-old, same-old*. But take a closer look around you, at our world and our culture. We hunger to make a difference. We want to stand for something. This might well be the strongest, most powerful, most universal human drive.

My *intent* (there's that dangerous word again) is not to hype matters so much as to strip away

- the veil of ennui, tedium, and dysfunction that often shrouds the workplace;
- the riot of distractions and urgencies of the moment that clamorously compete for our attention; and
- the complexities and intricacies of 21st-century living.

These combine to blind us to the big picture, but once gone, we can look at what remains: the high stakes in play in the 21st century and our part in the action. My intent is to distinguish true adventure from mere thrill seeking, and at the same time to highlight a circular connection between meaning and adventure: *Adventure* is *meaning* come to life in the moment. *Meaning* is *adventure* sustained over a lifetime.

This adventure and this meaning are accessible to you.

My *hope* (as opposed to intent) is that reading this book, and perhaps returning to it from time to time (re: handbook), will help you make a valuable and unique contribution to a better world. I hope that at the same time it will help you balance and prioritize your life's tasks and in so doing establish a rhythm of working on your particular piece of real-world living for extended periods. At intervals, remind yourself of the larger drama in which you play a part so that your work and efforts bear fruit.

So much for what the book *is*, or was intended to be. Here's what it's *not*.

Since the book is written for and involves everyone, it can't be a particularly scholarly work. Don't look for any special erudition; my apologies, but you'll be disappointed. Depending on your background and profession, you also may not find many *new* ideas. It's more likely what you read will crystallize what you've already been thinking for years or prompt you to articulate some superior insight. Nor will you find a lot of statistics or extensive references or footnotes. In part, through these omissions I'm acknowledging a trend that in itself is one of the book's points. Increasingly, thanks to our growing connectedness, everything we know or think we know will come from the collective, a swarm intelligence, as we swim around in the 21st-century's information soup. Going forward, whatever pride or sense of accomplishment you and I may attach to having been the first or only person to hold a particular thought or idea will prove misplaced or unprofitable.

And (please forgive me!) you might not even feel you've found many *right* ideas. I'm not trying to be particularly accurate. Worse, I've not succeeded in being wholly organized and logical—partly because the world is changing rapidly underneath our feet; partly because I'm still puzzling through many of these ideas as I write, and I expect to continue in that discovery long after this book is published; and partly because the ideas are so intertwined as to defy a truly satisfactory exposition. But also, for present purposes, it's far more important that the ideas here be stimulating than it is for them to be "right." As Charles Darwin once said: "False facts are highly injurious to the progress of science, for they often endure long; but false views, if supported by some evidence, do little harm, for everyone takes a salutary pleasure in proving their falseness." (*The Origins of Man*, Chapter 6)

Please, therefore, think of this book instead as an extended essay, or a collection of essays, or a set of conjectures at an early stage of development. *The book is scratchwork, a point of embarkation, not a destination. Any real utility in this book lies in what it prompts you to do after you put it down.*

You might be skeptical at this point. I fully understand. As you read further, your skepticism may even grow. The thinking you'll find here on issues of feeding a hungry world, slaking its thirst, or meeting needed energy requirements might seem out of step with much else that you and I read. Most scholarship suggests over and again that as individuals, nations, and a world we are failing to cope with these challenges. Experts give analysis and reasoning to show that we are losing ground with respect to vital particulars, and that over time we are destined to fail with respect to the grander enterprises. The message? *We're losing the battles and, what's worse, eventually we'll lose the war.*

Dour as these outlooks are, they tend to be too optimistic in one sense. They leave the impression that although we're going about our business on the planet in the wrong way, there is indeed a so-called right way, and if we but adopt it, we can live indefinitely along those (static) lines. (A popular example of this kind of wistful thinking: If only we would hit the CO_2 off switch, Earth's climate would stabilize.) Those subscribing to this roseate outlook call it sustainable development. The fact is, however, that sustainable development framed in this manner is an oxymoron. There is no such steady state available to us. We can't live sustainably so much as through a process of continuous innovation and change we can buy ourselves time.[1]

However, a main thrust of this book is that even though living successfully on the real world is far more challenging than most people realize or acknowledge, powerful new opportunities present themselves for doing just that. Promising trends are underway to understand both the planet we live on and our social behavior; the trends continue in policy, information technology and social networking, and our ideas of leadership. These developments are taking the world in directions that may make the 21st century mankind's best days so far. These trends are already in motion, however nascent, perhaps not quite so visible to us as we are caught up in the day-to-day turbulence of our jobs, family, and society. As a result, we're finding it hard to visualize ourselves succeeding, to see any clear path to a set of successful outcomes. If we're pessimistic, perhaps it's only because we're trying to solve *tomorrow's* problems with *yesterday's* tools, or because we take too grand a view of our capabilities as individuals and too impoverished a view of what we can accomplish as a collective.

There's one final piece to this book. Over the past century or so meteorologists have established a beachhead in the intellectual territory—new ways of doing business—that our larger society is poised to invade. Meteorologists

---

1. We might rightfully call ourselves the Scheherazade generation, after the spirited woman of Persian legend who survived for 1,000 nights in the presence of a king looking to behead her. By stringing together a succession of stories so fascinating he couldn't bear to leave a single one unfinished, she survived night after night to complete the last and start the next.

are accustomed to working on a nonlinear, complex problem—weather prediction—that defies and will always continue to defy any actual solution. Meteorologists have therefore been forced to embrace their limitations in a way that other physical scientists haven't, to settle for muddling through versus reaching actual solutions. They've also been forced to cooperate, to come to terms with their fundamental interdependence, in a way that other scientists have resisted. They have learned how to make progress by letting go of any aspiration to a truly comprehensive view and instead to make progress in small bits by living in the moment and thinking inside the box (although this sounds pejorative to today's ears). Throughout history, meteorologists have shared their findings with the broader public, but speaking the public's language, not using the argot of science. Meteorology leads neither to wealth nor to Nobel prizes. But meteorologists see daily how seemingly insignificant details in the weather can magnify over time to become major storms, and so it's natural for them to believe that each of us can through our thoughts and actions make a similar difference in the affairs and outcomes of a seven-billion-person world. Meteorologists know that it's possible to lead locally, from below.

This is the extraordinary part. Every aspect of this meteorological culture, mindset, and worldview can inform and equip 21st-century society. This meteorological attitude contains elements the whole world not only needs but, remarkably, is beginning to copy, though not necessarily by that label.

It is this kind of moment-by-moment, place-based, locally oriented approach to the 21st century that is the way forward for the human race, for good or evil. It's what's unfolding around us. We can dramatically improve our chances of successfully navigating the next 100 years if we get in tune with this reality; if we foster and become more purposeful about some of the trends underway; and if we give them a helpful nudge here and there.

So as a secondary element, I've made an effort to show this connection throughout the book. And because meteorologists make predictions, this book concludes with a few predictions about the future—what it will look like and how future generations might view us.

Chances are good—virtually 100 percent—that if you give the matter some thought, you'll find areas within your chosen profession or lot in life—mother, dad, schoolteacher, butcher, auto mechanic, farmer, carpenter, classics professor, clergy, nurse, politician, journalist, computer nerd, housekeeper, artist, county official, bus driver, subsistence farmer, refugee, government worker—that contain a culture and thought process that is similarly vital to the world's future.[2]

Please embrace that. Think big. Share your unique perspective with the rest of us. Today's IT and social networking give you the means to do that. Use what you find here as a point of departure for your own complementary, alternative, or superior vision. Please take the fullest measure of what Charles Darwin asserts is salutary pleasure due you for improving on, or proving wrong, these thoughts. Don't just talk to the rest of us. Listen. We all have a lot to teach each other.

While we're at it, perhaps we can pay homage to Henry David Thoreau urging us on (and in the process thinking like a meteorologist): "Go confidently in the direction of your dreams. Live the life you've imagined. As you simplify your life, the laws of the universe will be simpler."

---

2. This list reveals another of the book's limitations. Our future prospects depend in part on our comprehending the full diversity of cultures and approaches to problem-solving that are available only worldwide. Yet my experience of the world is limited and my perspective undisguised American. An Asian, African, or Latin American would offer a different perspective. I hope some will make the effort.

PROLOGUE

Once upon a time, hundreds of thousands of years ago, the very first men and women walked the earth. They were hunter-gatherers. They would move from campsite to campsite, following game as it migrated and picking fruits, grains, and vegetables in their season. For these nomads, every morning of their lives, at dawn's light, the first order of business was to build situational awareness (in other words, make a meteorological nowcast):[1] What's today's weather? Warm? Cold? Dry? Wet? Does it look to be changing? Is the cold season coming on? What do these weather signs portend for wild game, grains, berries, water, and our safety? Should we move on or can we afford to stay put one more day?

Everyone from the family elder to the youngest child thought about meteorological matters and thought as a meteorologist.

Thousands of years later, the nomadic lifestyle had given way to agriculture clustered around fixed villages and towns. The farmers and city dwellers—contemplating the salt-poisoned soils resulting from early, flawed attempts at irrigation, the newly discovered stresses and vulnerabilities of in-

---

1. A nowcast is like a forecast, except that instead of characterizing the future outlook it summarizes present conditions.

town living, and the rise of corrupt, oppressive nation-states—looked back and realized those diverse prior times and places actually had a single name.

Eden.

As civilization took root in the Fertile Crescent, it was accompanied by, and in like measure owed its existence to, a flowering of science and technology. The invention of the wheel transformed the movement of people and goods and prompted the development of the chariot—armament that was as potent at the time as the tanks and armored personnel carriers of today. Humans began to *mine* for metals and more. Harnessing fire fueled a transition from Neolithic culture to the Bronze Age and then the Iron Age. Mastery of clay and other materials led to the development of cookware, tools, and other utensils. The invention of money, and with it the counting board and the abacus, enabled and supported a new service industry: *trade*. From arithmetic came mathematics. Study of the night skies revealed regularity and predictability in stellar and planetary motions. To support agriculture and meet the needs of city dwellers for water, civilizations developed massive networks of aqueducts and irrigation ditches.

Only one scientific challenge stubbornly resisted human progress, despite the best efforts of mankind.

Meteorology.

The people of the period could still do no better than produce the same single-point nowcasts (*what is the weather like where I am now?*) that their nomadic ancestors had relied on. In fact, when they wanted to characterize the vast gulf between the capabilities, reach, and power of God and puny man, they chose weather to drive home the point.

We find this, for example, in the account of Job, one of the earliest writings making up the Bible.[2] Job is portrayed as one of the wealthiest men of his time, but God allows Satan to strip Job of his wealth, his children and grandchildren, and, finally, his health. Job is all but dead. That background tees up the main narrative, throughout which Job complains to his friends about the injustice of God's treatment. They offer cold comfort. Elihu, a young man in the party who has been silent throughout, grows increasingly

---

2. Job confronts us with one of the most important, age-old questions we have for God: Why do the innocent suffer? But our focus here is on another aspect of the story.

frustrated with the conversation. He searches for a question that will bring Job to his knees, one that will wake him up to the reality that he is but a feeble man who has no right to question the Maker of Heaven and Earth. Finally it hits him. He asks: "Hey, Job, can you forecast the weather?"[3]

Of course, Elihu puts it a little differently. Here's the text (Job 36:26–37:1–18, New International Version [NIV]):

> How great is God—beyond our understanding!
>> The number of his years is past finding out.
> He draws up the drops of water,
>> which distill as rain to the streams;
> the clouds pour down their moisture
>> and abundant showers fall on mankind.
> Who can understand how he spreads out the clouds,
>> how he thunders from his pavilion?
> See how he scatters his lightning about him,
>> bathing the depths of the sea.
> This is the way he governs the nations
>> and provides food in abundance.
> He fills his hands with lightning
>> and commands it to strike its mark.
> His thunder announces the coming storm;
>> even the cattle make known its approach.
>
> At this my heart pounds
>> and leaps from its place.
> Listen! Listen to the roar of his voice,
>> to the rumbling that comes from his mouth.
> He unleashes his lightning beneath the whole heaven
>> and sends it to the ends of the earth.
> After that comes the sound of his roar;
>> he thunders with his majestic voice.
> When his voice resounds,
>> he holds nothing back.

---

3. Recall that this question had a special bite back then. In Israel at that point in history, the penalty for false prophecy was death by stoning. Today's cultures are more forgiving of meteorologists.

God's voice thunders in marvelous ways;
    he does great things beyond our understanding.
He says to the snow, 'Fall on the earth,'
    and to the rain shower, 'Be a mighty downpour.'
So that everyone he has made may know his work,
    he stops all people from their labor.
The animals take cover;
    they remain in their dens.
The tempest comes out from its chamber,
    the cold from the driving winds.
The breath of God produces ice,
    and the broad waters become frozen.
He loads the clouds with moisture;
    he scatters his lightning through them.
At his direction they swirl around
    over the face of the whole earth
    to do whatever he commands them.
He brings the clouds to punish people,
    or to water his earth and show his love.
"Listen to this, Job;
    stop and consider God's wonders.
Do you know how God controls the clouds
    and makes his lightning flash?
Do you know how the clouds hang poised,
    those wonders of him who has perfect knowledge?
You who swelter in your clothes
    when the land lies hushed under the south wind,
can you join him in spreading out the skies,
    hard as a mirror of cast bronze?

Elihu's question throws Job for a loop. He ponders the notion. Predict the weather? No one can do that!

God Himself then piles on. He speaks out of the storm (Job 38:18–30):

Have you comprehended the vast expanses of the earth?
    Tell me, if you know all this.
What is the way to the abode of light?
    And where does darkness reside?

Can you take them to their places?

Do you know the paths to their dwellings?

Surely you know, for you were already born!

You have lived so many years!

"Have you entered the storehouses of the snow

or seen the storehouses of the hail,

which I reserve for times of trouble,

for days of war and battle?

What is the way to the place where the lightning is dispersed,

or the place where the east winds are scattered over the earth?

Who cuts a channel for the torrents of rain,

and a path for the thunderstorm,

to water a land where no one lives,

an uninhabited desert,

to satisfy a desolate wasteland

and make it sprout with grass?

Does the rain have a father?

Who fathers the drops of dew?

From whose womb comes the ice?

Who gives birth to the frost from the heavens

when the waters become hard as stone,

when the surface of the deep is frozen?

At this, Job throws in the towel. In the end, he's satisfied to let man be man and God be God, and he's content to leave his poignant, piercing, universal question of the ages hanging. He and God are reconciled. God admonishes Job's false friends, who had been pouring salt on his wounds, and restores Job's health, family, and fortunes.

Fast-forward another few thousand years to the mid-1800s. A young man by the name of Samuel Morse invents the Victorian Internet.[4]

At the time, they didn't call it that; instead, we know his invention as the telegraph. The telegraph and other IT of the day—the telephone (forerunner of the iPhone), the Victorian iPod (the phonograph), the Victorian cinema (the kinescope)—started a revolution that goes on today, a revolution that has and continues to transform every aspect of our lives.

Meteorology (finally!) benefited as well. Prior to the invention of the

---

4. In 1998, Tom Standage wrote a wonderful book by this title.

telegraph, meteorology hadn't changed much since the days of the hunter-gatherers. Postal services had hinted at a new possibility, which remained for the telegraph to exploit. When natural scientists of the time, Benjamin Franklin and Thomas Jefferson to name a famous few, corresponded, they discovered that storms and good weather apparently formed patterns, and that these patterns moved. For example, they would receive a letter from a friend in a distant location who had written about a storm, and they soon realized that between the time the letter had been written at the distant point and been mailed and arrived, there had been a very similar storm *here*.

The telegraph allowed them to remove any lingering doubt. When telegraph operators started to report prevailing weather conditions at their location as they began their shifts, and at regular intervals, it became possible to construct weather maps, which revealed these patterns and their motion and changes over time. This information networking, together with instruments such as thermometers and barometers and anemometers and wind vanes and rain gauges, gave birth to the profession of meteorology. Meteorologists could make nowcasts of the weather not just at a single point but over broad areas and, given enough time, over the entire globe. And using their instruments, but thanks primarily to a new tool for social networking, they could contemplate making forecasts.

That was the beginning.

1

INTRODUCTION

No man is an island,
Entire of itself.
Each is a piece of the continent,
A part of the main.
If a clod be washed away by the sea,
Europe is the less.
As well as if a promontory were.
As well as if a manor of thine own
Or of thine friend's were.
Each man's death diminishes me,
For I am involved in mankind.
Therefore, send not to know
For whom the bell tolls,
It tolls for thee. —*John Donne*

To be human is to be connected. We have always been and always will be. Our lives are interwoven with our relationships with family, neighborhood, and country. Poets like John Donne have long celebrated this. But today, that connection has been expanded. It is global. It defines not *humans* but *humanity*.

1.1. The Age of Interdependence (Nonlinearity)[1]

> Linear problems are all the same. Nonlinear problems are all different.
> —*Robert Hooke (1918–2003)*

My father, a mathematician, once shared this offhand remark, at a time when I was still in school. He certainly wasn't claiming this to be an original thought; it was simply an observation he made during whatever conversation we were having at the time. Ever since, I've reflected on the wisdom of this. Early on, for example, when I was finishing up my graduate work at the University of Chicago, I realized that in my thesis I'd stumbled on a small problem in an obscure corner of atmospheric science that had never been tackled before . . . and yet was linear. The mathematics was a breeze.

In a linear world, things happen more or less independently. They don't interact with each other. Their impacts are simply additive and separate.

Consider, for example, the dawn of mankind. There is a village or clan here, another there. Small nomadic groups are roaming over a vast land, making only sporadic contact with each other. The resources they consume are tiny compared with Earth's resource store, and largely renewable at that. In fact, the resources they consume are tiny compared with that consumed by other animal species. They take a fish from the sea, and the whales and other marine mammals don't notice. What happens to one clan or tribe in the Americas doesn't matter in any discernible way to another clan or tribe in Africa or Europe.

We came from this linear world. We're wholly adapted to it. This is our mindset. As far as mankind is concerned, our entire history has been lived in the Linear Age.

---

1. More likely than not, the age we live in—not unlike the Stone Age, or the Age of Exploration, or the Age of Reason—has a name. Such names usually become clear only in hindsight. But that shouldn't stop us from speculating and offering working titles, to be criticized and refined and made more eloquent. Here is the first of several candidate names that will be offered throughout the book (some original, others not). The material in this section was adapted in part from a post on LivingontheRealWorld dated January 25, 2013: "Our Age has a name . . . The Age of Nonlinearity."

Any such label is inherently flawed, and this one is no exception. For example, interdependence does not mean the same thing as nonlinear; neither perfectly captures the idea here. You also might be inclined to argue, and you'd be right, that interdependence and nonlinearity have always been with us. Touché! But let's go on, this won't be the last time I ask for your indulgence, forbearance, or forgiveness.

Until now.

In our present, nonlinear world, actions and events are no longer independent. They can no longer be simply summed. They have become *interdependent*. Agriculture and food consumption patterns the world over matter to local food supplies and prices everywhere. My dinner table here in Washington, D.C., reflects conditions in and contributions from Chile, Central America, Indonesia, and so on. In Indonesia, food prices and availability similarly mirror decisions and actions made in Europe, the United States, and elsewhere. What's on the dinner table in Jakarta, and what's missing, reflect that. The same holds true for water consumption; energy consumption; extraction of iron, copper, and rare earths; and more. For example, most societies have traditionally held drinkable water to be a public good and acted as if its consumption was nonrivalrous and nonexclusive. *My use of water for any purpose didn't matter. You still had plenty for your purposes, whatever they might have been.* Today, by contrast, we find ourselves confronted with stark choices about water use for agriculture, energy, industry, and public consumption.

This is particularly evident when it comes to extreme events, such as hurricanes, tornadoes, and cycles of flood and drought. Extreme events are by their very nature integrative—for example, take Hurricanes Katrina and Sandy, cases familiar to most Americans. These weren't simply extremes or even *natural* disasters. They didn't impact public safety and that alone. They damaged property. They disrupted business. They discombobulated transportation, not just locally but nationwide and internationally. Both were agricultural events; recall that the port of New Orleans is the embarkation point for most U.S. grains to the rest of the world. They were energy events. Gulf refineries, for example, are critical infrastructure for U.S. petroleum supplies. They were public health events, closing hospitals, and, in New Orleans's case, triggering a diaspora of trained healthcare professionals. They impacted the national finances.

This interdependence or nonlinearity is new to world affairs. It dates back no more than a century or so. Prior to that time, nonlinearities might have episodically surfaced at the local or regional level, but the impacts were largely local and tended to subside with time.[2]

Today, we find we are confronted with the interconnectedness of all things. Everywhere. All the time. At all levels of society. There's no decision

---

2. As acknowledged earlier, the nonlinearities have always been there, but so small as to be indiscernible.

we can make or action we can contemplate that isn't full of implications for many others—ultimately, for all of us.

It's not surprising, then, that climate change or natural hazards or food and energy production, or any other major issue for that matter, prompts seven billion people worldwide to proffer literally billions of overlapping coping strategies that draw in and involve every sector, every walk of life. Out of this babble of voices, given enough time, a resonant note will rise. Out of a seemingly chaotic, incoherent set of conversations and actions will emerge what history will look back on and in retrospect recognize as humanity's response.

On the ground, however, and lived out day by day instead of on some kind of fast-forward, the experience is tortuous and wrenching. The more we know (and seven billion people are busily accumulating a raft of experience), the more we realize that every action has impacts, and that the missteps and fumbles seem to outnumber the good bits. Analysis of our shortcomings is beyond us, but actually improving our performance in light of that analysis is orders of magnitude more challenging still. This makes us pessimistic and cranky. Then we start taking out our frustration on each other. *It's easier to decry your faults than to fix mine.*

Out of a seemingly chaotic, incoherent set of conversations and actions will emerge what history will look back on and in retrospect recognize as humanity's response.

So think of humanity as leaving behind the Linear Age and entering the Age of Nonlinearity.

Here's a forecast, based on something else my father said, back when I was an adolescent: "You're experiencing growing pains. You will grow out of this."

What my father and others of his generation meant was that all that teenage awkwardness stems from rapid physical growth, hormonal changes, growing awareness, and the clicking-in of a sense of responsibility. It was not the way things would always be. It was a stage.

So the forecast for the human race? We'll live in the Age of Interdependence (Nonlinearity) for a long time. But we—our present generation—will come to *master* it rather than be overwhelmed by it. The generations to follow will find it as natural as the Linear Age seemed to our ancestors.

And a lot more interesting.

1.2. The Book's Main Premise

In our 21st-century world, time is at a premium. So let's get right to it. We'll circle back to the details later, but here's the thesis of this book in a nutshell:

The world's peoples face major challenges as we start to navigate our way through a problematic 21st century. These problems—accessing needed natural resources, protecting the environment, and building resilience to natural hazards—are very real. They're complex and consequential, and they're threaded through one another; taken together (and they can't be separated), they can seem overwhelming. The fact is that they are intractable. But we've talked ourselves into feeling unduly pessimistic about our circumstances. We can't solve this threefold problem once and for all, but we can develop coping strategies. In a strange way, this is how it feels when things are going well. And they're going well in part because we're beginning to tackle our problems in much the same way and with something of the same mindset that meteorologists approach the challenge of weather prediction. If we will lay hold of this analogy and see efforts to forecast the weather as a microcosm of our larger human dilemma and as lodestone or sunstone pointing the way forward—if we will get intentional and strategic about our approach to 21st-century challenges—we'll not only be more effective but feel better about the human prospect at the same time. We'll discover that the 21st century is, in fact, history's sweet spot. We'll wake up to the reality that we're all major actors, not bit players or mere audience, in the unfolding drama. And we'll find a satisfying significance to our lives and work and relationships that we've hungered for but didn't believe attainable.

We'll discover that the 21st century is, in fact, history's sweet spot.

Let's dig a little deeper into some of those major challenges. To start, seven billion of us are consuming natural resources of all types in stupefying quantities. Remarkably, supplies seem to be holding, at least overall. But here and there we find shortages, especially in poorer, developing parts of the world. This picture holds true for all resources but most poignantly so for the basic necessities of life: water, food, and energy. The real worry is that the pace of this resource extraction may exceed rates that the land, rivers, seas, and natural ecosystems can support and sustain. Signs of environmental deterioration, and loss of habitat and biodiversity, are prevalent and growing in extent. In addition, we find ourselves vulnerable in new ways to extremes of nature. Disaster tolls are rising and spreading as we grow more reliant on critical infrastructure. So humanity faces a threefold problem: dealing with the earth as a resource, a victim, and a threat.

Our human response to all this, as individuals and as nations, leaves something to be desired. First and foremost, we individually seem to lack the necessary smarts to master the task comprehensively, in its entirety. Together we have to solve these three problems simultaneously, locally as well as everywhere across the globe. But that's proving too tough! Even when we band together, when we try to work institutionally or nationally, we're doing little better. It seems the best we do is to tackle at most one or two of the three aspects at any one time, to the detriment of the third. For example, to capture needed resources, we scar the landscape with strip mines and deforestation. We replace natural ecosystems with agriculture. But when and where we try to protect the environment we find ourselves incurring costs and putting foresters, fishermen, farmers, and others out of work. When we build dams and levees to protect against flood hazards, we belatedly discover we've endangered riparian wildlife (salmon in the U.S. Pacific Northwest, for example) and the way of life of indigenous people who had depended on them. This has the feel of an intractable mathematical problem but one with existential stakes:

Resource development. Environmental protection. Resilience to hazards. *Choose any two.*

The world's "have-nots" are quite simply overwhelmed by the day-to-day difficulties of finding water fit to drink, and feeding, clothing, and sheltering their children, all on a dollar a day or less. They often lack the education and skills necessary to cope with such challenges. In many cases, malnourishment from birth and the ravages of childhood disease have limited their own adult capacities and potential. By contrast, the world's "haves" are preoccupied and distracted. They are trying to navigate a frenetic, out-of-control 24/7 lifestyle in a virtual world of human construction that obscures underlying real-world conditions and trends. They find work competing with family responsibilities even as their finances are stretched to the breaking point. They seem numbed by the sheer scale of world poverty, see it as removed from them, simply push it from their minds, and do all too little to alleviate it. Joblessness is rampant across the globe, especially among the young, even as an aging demographic puts more pressure on that generation to support their elders. Some families have been unemployed for several generations, giving rise to subcultures with a *tradition* of joblessness. Nations underinvest in education and waste and misallocate funding for health care. Governments everywhere are spending more than they take in, especially as

economic growth has slowed. We compromise our future in order to pro-
long our comforts in the present. We consume resources; we don't replenish
them. We degrade the environment faster than we restore it. We are slow to
learn from experience. For example, after each disaster, we rebuild as before,
putting into place the same fragile infrastructure and flawed ways of doing
business, and condemning future generations to needlessly repetitive loss.

But all this is hard to see day to day. We're building up this bow wave of
problems at a point in history where, socially, we've lost touch with the real
world and our relationship to it. In the developed world, most of us work
in an artificial cocoon of pavement and air-conditioned buildings. We don't
experience our food as a product of rain and sunlight and soil; instead we
see it coming from the supermarket (or the restaurant kitchen). Water comes
from a tap. Energy comes from a flick of a switch or an ignition key. And
we're losing touch with even this constructed world as we enter a virtual
environment offered by IT. Today, two degrees of separation shield our view
of the real world we live on.

And come to think about it, we don't seem to be getting along that well
with each other. We're edgy, touchy, quick-tempered, sharp-tongued, and
contentious. Our discussions about what to do and why on almost any material subject quickly polarize, and our disputes lead to institutional and
governmental paralysis. We're quicker to criticize and find fault than we are
to work together to solve problems. Suspicion abounds, when instead what
we need most is trust. Ethnic, language, age, and even gender differences
divide us more than they should. Unrest, terrorism, and the threat of war
lie just beneath the surface nearly everywhere.

Perhaps it's a small wonder, then, that when we survey this scene, we are
overwhelmed by waves of pessimism. What optimism that remains is usually founded more on finding innovative ways to remain oblivious to these
realities rather than facing and dealing with them—addictions to drugs,
entertainment, and the accumulation of individual wealth are rampant.

The physical realities and social dysfunction combine to remind us that at
our core we have spiritual issues. All of us agree that we should do good, and
love, and forgive, and share. After all, we're *interdependent*. But we disagree
about the details, sometimes violently. We struggle to maintain contact with
the better side of our nature. The fact is that we don't like to acknowledge
this dimension of our lives so much.

In the face of this panoply of problems, we need to lay stronger hold on
realism; and we greatly desire to grow more *hopeful* in the process. To be
more precise: We don't want a delusional hope but rather long to discover

that reality has a bright side. And that bright side means solutions to the world's ills that are *quick, cheap,* and *actually work.*

A tall order.

But there's help. And this help is available from what might at first blush seem a most unlikely quarter.

1.3. Meteorologists

This is not about *who* meteorologists are. As individuals, meteorologists are pretty much the same as everyone else. Some are better-looking than others. Some are smarter. Some are really likeable, others maybe not so much. Some are positive in their outlook, others less so. It probably can't be argued they're a particularly spiritual or even well-adjusted group.

And it's not just about *what* meteorologists do. It's not simply that meteorologists forecast the weather, provide climate outlooks, and predict rain and snowfall. It's not that these services are used by agribusiness, energy utilities, or water-resource managers, though all this is so. What meteorologists think about and do is important, but that's not the only reason meteorologists offer hope to all mankind.

Instead, it is the *way* meteorologists are trained and inclined to think, and the *way* they approach problems, that is potentially far more valuable to society. It's the *culture* of meteorology, the *ideals* of the profession, and the *methodology* of the community that matters. It's *how* meteorologists approach their work that offers us something.

This starts with attitude. When it comes to weather forecasting, meteorologists are in way over their heads. Instead of trying to hide or deny the fact, they know and embrace it. They're realistic. They know that their nowcasts (i.e., their characterizations of present or initial conditions . . .) lack detail and contain errors, and that over any forecast period these will quickly grow with time, making any attempts to forecast very far into the future essentially worthless. They know that the equations on which they base their forecasts are only simplified approximations to reality—and worse, that even these approximate equations are absolutely intractable. They will not allow a closed solution.

Contrast this with physicists. Early on, Isaac Newton (1642–1727) developed his laws of motion and inferred the inverse-square law of gravitational attraction. He was able to solve that equation analytically; that is, he wrote a simple formula that foretold where two bodies in each other's gravitational field would be on out to the end of time. (It isn't quite that straightforward in real life, but almost so.) From that point on, physicists have been able to predict the movement of planets, stars, and other large astronomical bodies. They can list the dates and locations for solar and lunar eclipses out thousands of years into the future. They know when the comets are coming back. They know how close all the countless asteroids around us will come to Earth for centuries in advance. They can predict the Higgs boson should be there and then build big, expensive machines for the search; it takes a while, but they find it. Physicists have swagger, and rightfully so.

By contrast, the joke is that meteorologists struggle to tell you what'll happen in the next five minutes.

Meteorologists could well have given up. Thrown in the towel. No one would blame them. But meteorologists' attitudes also include "no sense of shame." With no embarrassment or apology, they've put up with all the jokes[3] and they've settled for their fuzzy nowcasts. They've replaced the insoluble equations with a bunch of arithmetic—simple sums and differences and multiplication and division that a sixth-grader could do—and called it close enough. (The only difference is that a sixth-grader doesn't have the speed or the patience for the millions of computations, so meteorologists in recent years—the moment they could—have relegated the task to computers.) Meteorologists run their calculations out a few model hours or a few days, and when the calculations clearly no longer bear any correspondence to reality or retain any utility, they abandon the exercise and do another fuzzy nowcast and start over and redo the new arithmetic, which will (sort of) work for the next few hours or days.

To an outsider, this might not look much different from a hamster running the treadmill.

But as meteorologists have thought and fretted and fussed over the process, as they've contemplated what went wrong in each prior forecast and how they might do better the next time, they've made a lot of progress.

For example, the year 2012 produced a striking example of how far the field has come. Hurricane Sandy, which formed in the Caribbean, threatened not only island nations but the Atlantic coast of the United States. Normally, hurricanes such as this one, as they move from the tropics northward toward the poles, encounter midlatitude westerly winds, and are carried harmlessly

---

3. In fact, the American Meteorological Society has just published a new collection of meteorological humor: *Partly to Mostly Funny: The Ultimate Weather Joke Book*, edited by Jon Malay, a former AMS president. Q: Where did the meteorologist stop for a drink on the way home from a long day at work? A: The nearest isobar! Buy it. You know you want to!

out to sea. This time, numerical models showed that instead Sandy would take a stutter-step to the west, making landfall over New Jersey, and then move through the interior northeastern United States. It would trigger major coastal flooding; inland, some areas would experience paralyzing snowfalls. As a recently as a few years ago, neither the observations (from satellites and weather radars and aircraft) nor the models would have allowed us to anticipate this move or inspired the confidence needed to trigger an unprecedented mobilization of the region. But in the face of Hurricane Sandy, local, state, and federal governments started moving their people and their assets well in advance of the storm's landfall. Insurance companies and bottled-water distributors and retailers from Home Depot to Walmart and Target did the same. So did the airlines and Amtrak and truckers and the electrical utilities.

This multifaceted mass mobilization based on meteorological prediction was a major accomplishment. But here's the eye-opening part. To study meteorological methods is to realize that they might be carried over from that field and successfully applied to those seemingly intractable real-world problems (world hunger, disease, poverty, terrorism, loss of biodiversity) that have us all feeling pessimistic. And, what's more, to look around—to read newspapers and books and to sample broadcast and other media— is to see nascent, emerging efforts just getting underway along precisely these lines. To navigate our way to a high quality of life in 2100 or thereabouts, we do not have to ask that society do a 180-degree turn; the promising developments are already being embraced. All that's needed is for the world's peoples to recognize what's afoot, join in, get slightly more strategic and intentional about things, and speed up the learning and the adjustments a bit.

But meteorologists have more to offer than just insights into the physical world. They also have social skills. Meteorology, unlike other branches of science, is inherently a communal endeavor. It's impossible to forecast weather a week in advance in Washington, D.C., unless Chinese and Russian meteorologists, and many others, share information on their prevailing weather conditions right now. For its one-week forecast, Europe needs the same help from us. And so on. The history is that meteorology has been a cooperative endeavor from the start. And the cooperation has so far been aimed more at benefiting the public and the world than benefiting meteorologists themselves. So, whether meteorologists self-select for entry into the field (it's overwhelmingly populated by men and women who've wanted to forecast weather since they were 10 years old) or whether they're shaped or encouraged or weeded out based on their education and training and the nature of the workplace, the community has a different feel from, say, physics. Or law. Or finance.

But it is meteorologists' attitude of *hope* that may prove far more important still. Meteorologists have always been convinced that their skills and capabilities are constantly on the rise, and that if those advances can be put to use they will prove to be of great practical help to others. More than a century of progress suggests this optimism is warranted.

This book unpacks these ideas. We start with some physical realities. To many, these may seem obvious on their face. But as individuals and as a society, we lean more toward ignoring them and their implications than responding to them.(Social realities are also important to this book, but we're going to postpone them until Chapter 3.)

LIVING ON THE (PHYSICAL) REAL WORLD

2

Reality is that which, when you stop believing in it, refuses to go away. —*Philip K. Dick, in* How to Build a Universe That Doesn't Fall Apart Two Days Later

The one dimension of the reality that you and I face daily that appears least flexible is the physical. It's the vital starting point for living on the real world. Here are some examples: Two bodies can't occupy the same space at the same time; momentum, energy, and mass are conserved; for every action there's an equal and opposite reaction; and so on. Jump off a cliff? Run full tilt at a wall? Chances are good you won't like the result. Physical realities of this type bound our challenge of living on the real world.

Because these physical realities have been ever-present and have shaped all of human experience to date, we rarely think about them consciously. But the implications they hold and the constraints and challenges they pose to today's interdependent society are different from what they implied to the world's smaller, more dispersed, subsistence-level populations of the past. They therefore merit a second look, a reexamination.

2.1. We're Earthlings, 'Til Death Do Us Part

The first physical reality? Our future is here. We're not going anywhere soon.

In 1975 or thereabouts, I was in my early 30s and living in Boulder.

During a visit with my mathematician father in Pittsburgh I enthusiastically reported that the University of Colorado had just constructed a new planetarium on campus, next to the astronomical observatory. To engage the public, they built a scale model of the solar system outside. They had a ball about the size of a basketball that represented the sun, and several feet away was a bb pellet, representing Mercury. Some distance farther was a pea-sized sphere: Venus. Another pea-size sphere representing Earth was next, and so on, out to the small speck representing Pluto (it was still considered a planet in those days), about a quarter-mile away, way off by the CU Engineering Center. "And," with a flourish, I told my Dad, "the final plaque says that to scale, the nearest star would not be in Boulder, but in Harvard Yard, in Cambridge, Massachusetts, 2,000 miles away!"

"Yes," my Dad answered, without batting an eye. "That's why astronomy is so uninteresting."

This from a techie, a science junkie—a member of AAAS, a fellow of the American Statistical Association, himself the grandson of a 19th-century astronomer. How could he say that? Because in his view, human destiny would be played out here on Earth. Taken further, we could be curious about the larger universe, but we needed to be serious about our home planet.

Our priorities needed to concentrate on this home ground—the Cold War, population increase and how to feed the hungry, the economy, education, the environment, and more. He saw these problems as pressing and as having immediate consequences and deadlines. When and if we met these challenges, there would still be time enough for exploration of the solar system and the universe and conjecture about its origins. The stars weren't going anywhere. The galaxies would still be around. The problems here were the only ones with a ticking clock. He didn't see how any other mindset could be realistic.

Do you agree? In that 1970s conversation, was my Dad simply being realistic? Was he right? Are we held here?

How about the colonization of space, should, say, Earth become uninhabitable? By dint of special effort we periodically launch a handful of people into a temporary orbit. By more effort still we could return to the Moon (at this writing the Chinese are contemplating such a visit in a decade or so) but not in a way that would enable a number of us to stay there indefinitely or for us to become in any sense of the word independently self-sustaining.

And as for human colonization of the planets, that seems even less likely (at least for now). Take Mars, our nearest neighbor. At its closest, Mars is

30,000,000 miles away, more than 100 times as far as the Moon. The Moon is just a few days' journey distant; getting to Mars with today's technology, or any engineering on the horizon, will take months. And when we get there, we'll find that Mars's atmosphere is about 1 percent the density of ours—essentially a vacuum—and mostly carbon dioxide. Nobody's living on Mars except in spacecraft-like shelter. No large numbers will be living there over any historical interval that we care about. We have yet to prove that we can get there and get back. (We're likely to give it a try circa 2030–2040.)

Space, as my Dad observed, is largely empty (physicists tell us different these days, but the distinctions between a vacuum and dark matter and energy are a little over most of our heads). It is trillions of miles to the *nearest* star, let alone one of the other hundred billion in our galaxy. The others are *much* farther distant. Astronomers tell us that there are so-called Goldilocks planets out there: Planets that are not too hot and not too cold but rather where the combination of gravity and temperature and atmospheric composition may be just about right for us and where water might exist in liquid form. But they're rare and thousands of light-years from here.

An update: For the immediate present, human space exploration feels more like something we *used* to do. We gave it a try, but it proved difficult and unrelentingly expensive and we've pulled back for a while. We've hit the "pause" button for now.

2.2. The Second Physical Reality? We're Confined to the *Skin* of Earth

The word *confined* seems negative, and maybe we'll get back to that and reframe it, but before we do, let's look at our experience here on Earth a little more thoroughly. It turns out on closer inspection that we by no means have free rein around the place. We don't live *in* the real world so much as we live *on* it.

Hence the book's title.

Our home planet—that pea in the CU Planetarium exhibit—is in reality some 8,000 miles in diameter and 24,000 miles around. But human activity for the most part is confined to the *skin* of Earth: the very thinnest sliver right where the earth and its waters meet the sky. How thick is this skin? There's no sharp boundary here. But think in terms of a narrow region between 40,000 feet above Earth's surface—airplane flight altitude—and maybe a few thousand feet below that same surface, which are the depths we mine. How numerous is the flying public? Maybe 100 to 200 people per plane, say, on 30,000 two- to three-hour flights per day. At any given moment, no more than 300,000 to 500,000 of us out of seven billion are airborne—one out of every 20,000 people or so.

How about underground? How many are making a living working in subterranean (as opposed to pit) mines? Perhaps more people than are flying, maybe a few million, but still only one person in 1,000. Could we all start burrowing underground in larger numbers any time soon and maybe even live out our lives there? Not easily. It turns out that rock temperatures increase maybe 25°C to 30° C for every kilometer of depth (that's maybe 100°F for every mile down); the deepest mines in the world as of this writing extend no more than 4 km (2.6 miles) beneath the surface. The heat is oppressive, pressures are enormous, and accidents are a way of life—and of death.

Realistically, all of us are going to be spending essentially our entire lives within a few feet above or below the (modified)[1] Earth's surface. The same thing is true of virtually all plant and animal life. Whatever impact we're having on planetary processes—on the environment and the landscape and the habitat and the wildlife—is confined to the surface. We may be engaged in what appear to us to be life-and-death matters of the greatest import, but from the standpoint of Earth, we and our activities represent no more than some mild (and probably temporary) sort of planetary skin disorder.[2]

2.3. Unpredictable

The third physical reality? It's more difficult to predict what will happen in the next five minutes here on Earth's skin than it is to predict where Earth itself will be in 100 years.

Or 1,000 years. Or maybe 1,000,000. That's because Earth's atmosphere and oceans are examples of what scientists call chaotic systems. Small details in the conditions at any one time can grow into major disturbances, sometimes very quickly. This sensitivity to initial conditions has come to be known as the butterfly effect. The idea is that some unremarkable detail of an atmospheric initial condition, say, a butterfly flapping its wings in Africa, might set into motion a chain of disturbances of growing scale that might result in a hurricane hitting the Americas some time later. (Or not; in another setting, another butterfly's contribution might be to suppress a storm

---

1. For example, we might be in a skyscraper, a metropolitan area subway, or a pit mine.

2. In that light, the book's subtitle is a little over the top. Again, if you take offense, my apologies.

that otherwise would have developed.) It's this character that prevents our forecasting weather specifics years in advance.[3]

2.4. Extremes

The fourth physical reality? Earth accomplishes much of its business through extreme events.

Here are just a few examples. Geologists make much of continental drift: the slow movement of the continents and the formation of oceans and mountain ranges driven by circulation of magma deep in Earth's interior. On Earth's crust, however, that slow drift manifests itself as a succession of cataclysmic earthquakes. The inch-per-century average motions are really made up of seismic upheavals. The great Tohoku earthquake of March 11, 2011, created a tsunami over 100 feet high at some locations, moved the entire main island of Japan 25 feet, and shifted Earth's axis of rotation an inch or two, all in a matter of seconds. In other locations, similar shifts trigger massive volcanic eruptions, such as Iceland's Eyjafjallajökull in 2010, the Philippines' Pinatubo in 1991, and Indonesia's Krakatoa in 1883 and Tambora in 1815.

Earth's climate is as mild as it is because great atmospheric and ocean circulations transport heat input from the equator to the poles. But these circulations do only about half the job. The other half is accomplished by the fierce storms and eddies that ride atop these grand, slow currents. We have a saying that *when it rains, it pours*, because most everywhere the world over, rain doesn't occur in any continuous drizzle so much as it occurs through episodes of flood and drought. In the United States, Seattle has an annual average rainfall roughly comparable to that of Dallas, despite the fact the latter is sunny and dry most days. In many tropical locations, hurricanes and typhoons deliver significant fractions of the rainfall averaged over an entire year. In Colorado, the Big Thompson Canyon flood of 1976 resulted from an average year's worth of rain falling on the watershed in eight hours.

Thus extremes of nature are not momentary suspensions of the natural order. To the contrary, they embody the way things are.

(Note especially that extremes aren't in and of themselves a bad thing. Extremes aren't necessarily hazardous. It's our ways of doing business that make them so. To reemphasize, extremes are nature's way of doing business. But *disasters*—disruptions of entire communities that persist after the ex-

---

3. Hold this thought. We'll be circling back to it several times in this book, in different contexts.

treme has come and gone and that exceed a community's ability to recover unaided—are the product of human decisions. Risky land use. Deficient building codes. Reliance on fragile infrastructure. Tolerance of chronic poverty. And so on. More on this in Chapters 3 and 8.)

2.5. Entropy

The fifth physical reality? Disorder in the universe is always increasing.

This may be the harshest, least forgiving, most unyielding reality of all. According to the laws of physics, things are growing more disorderly over time. Physical scientists have a measure for this disorder. They call it *entropy*. But for purposes here, it suffices to know that this is merely a mysterious label for something most everybody knows well from experience. It's easier to squeeze the toothpaste out of the tube than coax it back in. It's easier to break a pane of glass than reassemble the shards into that original pane. It's easier to burn a gallon of gasoline than to capture all the particulates and molecules released by that combustion and reconstitute the gasoline. *All the king's horses, and all the king's men, couldn't put Humpty Dumpty together again.* It's easier to melt an ice cube while dousing a flame than it is to restore both the ice cube and the flame from what's left at the end. It's easier to injure a social relationship than repair it. It's easier to kill than to bring back to life, to resurrect. It's this encompassing reality that underlies the choice we lamented upon earlier:

Resource development. Environmental protection. Resilience to hazards. Choose any two.

We lack (and rather urgently need) a holistic approach to the real world. In fact, here's the ideal: In every decision and action, large and small, we take fully into account that the world is simultaneously a resource, a victim, and a threat.

Today we call this ideal sustainable development.

2.6. Sustainability

Sustainability is an oxymoron.[4]

---

4. *Oxymoron* might seem a harsh label for an important aspirational goal. Perhaps we should call it instead the Holy Grail of real-world living. Recall that (in medieval legend) the Holy Grail was the bowl used by Jesus at the Last Supper. It was allegedly brought to Britain by Joseph of Arimathea, where it became the quest of many knights. That quest, however, was always futile.

The Gro Brundtland report, "Our Common Future," which was published in 1987, provides both a definition and a goal: "Sustainable development is development that meets the needs of the present without compromising the ability of future generations to meet their own needs."

The Brundtland formulation has been refined over the past quarter-century in reports and studies, but it has, generally speaking, held up rather well. The idea as expressed is simple and compelling. We can all grasp it. But the simplicity, though seductive, is deceptive. The idea behind sustainable development and the execution of it both pose problems.

To better see this, let's look at two or three real-world challenges to sustainable development.

First, any use of a nonrenewable resource requires a continuing process of innovation. We can't stand still, because per capita appetite for the resource is ongoing, our numbers are growing and our recycling is imperfect. Alternatively, and a lot more realistically, what we're really doing is no more than "buying time." Pick any activity—resource extraction, resource use, waste management, financial transactions, health care, education—you name it. Suppose we were able (and this is indeed a hypothetical) to calculate how long we could continue to indulge in that activity in the present way before hitting some limit.

The *Bulletin of Atomic Scientists* doomsday clock is a very specific but wonderful example of how this might look. For decades this publication (now available only online) has attempted to portray how close the world is coming to catastrophic destruction through a graphic image of a clock indicating that it's only a few minutes before midnight. During the Cold War, the focus was on the threat of nuclear war. In more recent years, the clock reflects an expert view of the combined threat represented by nuclear war, climate change, and bio-security trends. The precise time displayed moves backward or forward from time to time reflecting any current easing or hardening of these risks. Subjective to be sure, but a great way to capture attention and prompt thought and discussion and, on occasion, action.

Such a clock, then applied to real-world circumstances generally, would make it clear that no activity is ever sustainable, because future generations would face a shorter time during which they could indulge in that same activity. Unless . . . we innovate our way to greater efficiency.

Here's a hypothetical example. Consider the use of some nonrenewable resource (take your favorite rare earth or other strategic mineral), with known exploitable reserves that would last a century. If each year we became something like 1 percent more efficient in our use of that mineral, or 1 percent more efficient in our ore-extraction methods so that we could increase the lifetime and size of the reserves, we could draw on that source forever. (Unless, that is, the supplier held a near-monopoly position and decided to punish one or more customers for real or imagined bad behavior. In 2012, China did just that with the Japanese, who had the effrontery to jail a Chinese fishing boat captain for trawling in Japanese waters.) Note, however, that there's no getting off this treadmill. We have to keep the innovations coming, year-by-year or fall behind the sustainability curve.

Unless . . . we could prospect more cleverly, or find some renewable-resource substitute or perhaps a wholly new product or service that would obviate our prior dependence on the rare Earth element, or approach 100 percent recycling of the material, or you get the idea. There are a lot of ways to go. For this reason, economists tell us we can relax. Pricing will drive the needed innovation, which is in inexhaustible supply. We can only hope they're right.

And we can innovate in ways additional to science and technology. Briefly, in the case of the doomsday clock, innovations such as nonproliferation treaties, the reining in of rogue states, or the resolution of disputes all serve to move back the hand of the clock and buy time.

There's an additional sustainability challenge associated with resource extraction: the environmental despoliation that is part of the process. Take minerals. Strip-mining scars the landscape. Underground mines produce huge amounts of tailings and often prove vulnerable to slumping. One approach we've adopted to this problem in the United States is "out of sight, out of mind." We've put regulatory restrictions on such extraction. This has had the effect of moving such operations overseas. Our balance of payments in trade may be negative; we import more goods than we export. But we export more environmental damage and risk than we import. We then pat ourselves on the back for cleaning up the environment. (Only belatedly, if at all, do we notice the deleterious effects of this abroad. Then we sometimes act as international scolds. *Why are our supplier-nations allowing such environmental degradation? Savaging their landscape? Polluting air and water? Destroying biodiversity? Why can't they get on top of this the way we have? And by the way, if they refuse to do so, isn't that unfair trade?*)

So far the discussion has focused on sustainability of the supply side. What about sustainability of the use? Any use or consumption of a resource leads to waste, and with this another whole series of issues arise. Where and how to dispose of the waste? "Dispose." Another oxymoron? We don't dispose of waste so much as we move it around. And the scale of that waste removal and the distances involved have grown. New York City waste has created new topographic features in the outskirts of the city. Waste that was once trucked to New Jersey is now trucked to western Pennsylvania. (*Coming soon to a rural, undeveloped area near you.*) These problems are particularly pronounced as population and the resource use per capita both rise. As we've seen, these have been major realities of the past century or so.

The concern has led to recycling: In many respects, and for many materials, this is a huge improvement. It buys us more time. But it's still imperfect. We don't recover all of any given material (plastics, metals, paper, etc.). We also don't recover all materials, and so recycling is another arena that demands continuous innovation. We constantly have to add new dimensions. Look around you. You're seeing this in your lifetime. Perhaps the greatest example is the recycling of all of our new gadgets—computers and peripherals, cell phones, and batteries, to name a few. (We even have a name for it: *techno-litter.*) And many toxic wastes that you and I used to "dispose of" willy-nilly, such as paints and a wide range of petroleum products, must now be handled in a controlled way. In Europe, the full life-cycle costs of many such items, including the costs of their disposal, are folded into the purchase price. Soon that will be the case here.

The elephant in this particular room is energy, especially fossil fuels. They are not converted into products but instead truly consumed. The costs of extraction are rising. These costs do not fully account for the environmental degradation (fine particulates, ozone, oxides of nitrogen, and a range of hydrocarbons in urban smog; Houston, we have a problem) at the point of use, or at the extraction sites (do people still remember the BP oil spill, or has that memory already faded?). These costs are externalities. Use of fossil fuels has become so widespread as to introduce global *The elephant in this particular room is energy, especially fossil fuels. They are not converted into products but instead truly consumed.* consequences (climate change and attendant effects). Nations, corporations, and individuals are constantly searching for additional reserves, introducing fuel economies, turning to nuclear energy, and developing renewable energy alternatives.

So, to rephrase the Brundtland definition: Buying time—meeting today's needs while minimally compromising the prospects for future generations—is the defining, and ongoing, challenge for every generation.

Mitigating circumstances are at work even here. It turns out that disorder of a closed system doesn't always have to increase. It can remain the same, so long as what physicists call "irreversible processes" don't occur. Picture a heavy rain's effects on the lowlands of a watershed. These may temporarily flood, but if the waters recede before the vegetation has started to die, or before the bridge crossing the river has begun to fail, then the disorder of the system hasn't really increased so much. But if the floodwaters persist, and the grasses and the trees die so that after the waters recede they fall prey to parasites, and if the bridge and the road wash out, then disorder has reared its ugly head. In the atmosphere, it's possible to warm air by compressing it. To some approximation, this is what happens when air sinks, as when it flows down a mountainside, say. Entropy doesn't increase in this process, except in the small layer of air immediately adjacent to the surface of the mountain. But warm the air by setting off a firecracker? That process is irreversible, and you've raised its entropy.

Our Sun provides an energy source we can use to our advantage. If we can develop the technologies to capture more of that solar energy than we do currently, if we're sufficiently smart about how we harness that energy to capture the resources we need, if we can maximize our use of renewables and recycle the nonrenewables, then we can buy time for ourselves—perhaps as much as a billion years if we're clever and if we continue to innovate.[5]

All of this leads to the first of several proposals in this book—a series of initiatives.

Real-World Living Initiative #1. In addition to speaking of sustainable development, let's initiate a number of analyses with respect to the needed natural resources that estimate how much time we have. Let's set up a limits clock, analogous to that Bulletin of Atomic Sciences *doomsday clock, for each resource. Let's update those clocks periodically, celebrate those occasions when*

---

5. An ecologist colleague tartly reminded me upon reading a draft of this chapter of another biological/physical reality: Most species only survive a million years or so, and based on that timing human beings may have pretty much run out their string. To which I can only express the hope that in this respect as well as others (considered in later chapters), we should consider the ant and be wise. So far as we can tell, ants have been around for 100 million years and are still going strong. Some have estimated they make up perhaps as much as 15 percent to 25 percent of terrestrial animal biomass. This book's focus is on a much shorter time scale: navigating the remainder of this century in such a way that our descendants can transit to the next one.

we buy ourselves some time, and focus our attention on those areas where time is short.[6]

2.7. Recap

This, then, is a partial, merely illustrative sample of physical realities. Earth does its business through extremes. The details of the atmosphere and ocean at any moment drive radically different outcomes over time. We're confined to the skin of Earth, and we're stuck here on Earth until further notice. There's no such thing as sustainable development, in the way most of us picture it; all we can do is buy time (but historically we've been pretty successful at this, and we can get even more proficient).

With this background, we are ready to look at another, parallel universe of realities: social realities. That's the subject of Chapter 3.

---

6. In a way, the commodities markets accomplish precisely this. They're especially insightful because they automatically include an element of crowdsourcing. They capture the thinking of many experts worldwide. But they also reflect short-term imbalances in supply and demand (hours, days, weeks, months) versus the outlooks over decades or centuries of interest here.

3

LIVING ON THE REAL WORLD (SOCIAL REALITY)

At the start of 1986 I was running a 200-person, 20-million-dollar-a-year National Oceanic and Atmospheric Administration (NOAA) laboratory in Boulder, Colorado. Our work spanned climate, meso-meteorology, and weather modification research, as well as systems development and rapid prototyping for decision support in weather. It was all physical science and engineering, all the time.

My career, and my perspective, were about to change.

Frank Press, a seismologist and former science advisor to President Clinton, was then president of the National Academy of Science (NAS). For several years he'd been working with counterparts across the world, trying to establish a United Nations International Decade for Natural Disaster Reduction (IDNDR). He felt an NAS National Research Council report was a necessary start but had been disappointed with an earlier result. He formed a new committee, which was supposed to have included Dr. Joseph H. Golden, a meteorologist with an extraordinary history. Among other achievements, Golden founded the Tornado Intercept Project in Oklahoma, which has since given rise to today's widespread tornado chasing, both for research purposes and as a form of eco-tourism.

For some reason, Golden stepped back from participating. Through the ensuing series of coincidences that so often shape our lives, I found myself appointed the committee's token meteorologist in his stead.

I knew nothing about disaster reduction.

Worse yet, I thought I did. After all, wasn't it obvious that the high winds and storm surge from hurricanes or the ground shaking from earthquakes cause the damage? What else was there to know?

The first committee meeting in Washington, D.C., was a revelation. The room was full of engineers, sociologists, emergency managers, and leaders of relief organizations. Geophysical scientists were in the minority, hardly to be found. Through the committee work over the next year I first learned that extremes are nature's way of doing business. But disasters—disruptions of entire communities that persist after the extreme has come and gone, and that exceed the communities' ability to recover unaided—are a human construct. They are the result of poor land use, inadequate building codes, poverty and other pre-existing social inequities and vulnerabilities, lack of preparedness, poor emergency response, and more.

I had to face it. In my obsessive focus on physics, I wasn't playing with a full deck.

3.1. Through the (Social) Looking Glass

I'd confronted this problem before—playing with a partial deck, that is. Meteorologists know that in specifying initial conditions for an atmospheric forecast, it's not enough to measure *some* of the variables, say, the three components of the wind: east–west, north–south, up- and downdrafts. It's necessary to measure *all* the variables, including the winds as well as the temperature, pressure, and humidity.[1] Without knowing *all* the initial meteorological conditions, it would be impossible to predict what happens next. There's no use cursing this particular darkness. It's the way it is. Successful meteorologists embrace this rather than flinch from it.

In the same way, there are social realities in addition to the physical that seven billion people must acknowledge if we're to navigate our way safely and effectively through the 21st century and live well on the real world.

In a nutshell, here are a few that are especially pertinent: (1) social realities are quite different from physical realities. They're not amenable to mathemati-

---

1. We'll be returning to this idea again and again throughout the book.

cal laws,[2] but that doesn't mean that there are no rules or that those rules can be ignored. (2) Powerful social changes are underway: the rapid increase in the number of humans, the population concentration into (mega)cities, the increase in resource use per capita, the globalization of the economy, and more. But underlying and accompanying all these trends are (3) a distancing of most people from direct contact with the real world on which we all depend and (4) a concurrent rise in the complexity of human affairs. Against the backdrop of this change, (5) human *nature* seems to be remaining constant. We have an inflated idea of our physical and mental capacities; we feel overly "lucky/ bold" in the face of some risks and excessively "pessimistic/timid" in the face of others. We steadfastly remain more inclined to act in our apparent short-term self-interest than in our long-term self-interest, and we are more likely to act to our perceived advantage rather than the best interests of others.

As discussed in Chapter 2, the physical challenges alone make for a hard slog. But here's a dirty little secret that most of us in physics and related fields know and keep to ourselves. The principles of mathematics and physics that seem so difficult to so many are actually quite elementary. It's far easier to deal with an electron in a potential well or even with the conundrum of particle-wave duality than it is to interact with another human being. The physical science thought process is rudimentary—the simplest pure logic. And mathematics and experiments are there to raise red flags when you've gone off course. They function like guard rails at bowling alleys that keep little kids (and the occasional grandparent) from throwing gutter balls or like the training wheels on a bicycle. So, if you happen to be a scientist, be aware that the show of respect that comes when people learn you're a mathematician or physicist or chemist isn't really deserved. And likely if our social skills were more developed, we physical scientists might realize that not all that fawning and respect is genuine.

---

2. Physicists and biologists alike might choose to quibble with this. For example, the great biologist Edward O. Wilson, in a wonderful little book titled *Letters to a Young Scientist* (Liveright 2013), asserts two fundamentals on which biology rests: "(1) all entities and processes of life are obedient to the laws of physics and chemistry, and (2) all evolution (subject to a few caveats) is due to natural selection." He would undoubtedly include the evolution of behavior under this umbrella. But at current levels of physics and chemistry, to discuss social behavior in terms of such laws does not lead to useful outcomes. And to wait for natural selection to make us more fit for living in harmony with each other may require time we don't have.

The truth is that when it comes to addressing our three requirements for living successfully on the real world—garnering resources, preserving the environment and ecosystems, and protecting ourselves from hazards—physical limits and technological challenges pale in comparison to the social problems we face. You'll hear natural scientists say, for example, that climate change poses a societal challenge because of its slow onset. But social scientists might make a case for the opposite: They recognize that climate change is rapid onset compared with the time required for seven billion people to agree on what to do about it. The world's rich, living in developed countries, want the (rapidly) developing world (China, India, and the rest) to forego fossil-fuel use in favor of relatively expensive renewables. The developing world asks for compensation; they see the current warming trend as the consequence of the developed nations' past mistakes and environmental abuses. The impasse is immediate—and so far, enduring.

Often social measures might in principle cost next to nothing to address. In fact, they might actually generate resources, create margin, and even buy time. The bad news is that, as with meteorology itself, social challenges have remained stubbornly intractable throughout most of human experience. And problems framed in these terms look quite different than when framed as physical-science-and-technology barriers.

To appreciate this, let's stick for the moment with the climate change example. Any atmospheric scientist will tell you: The science is in, and the science is clear. Greenhouse gases—carbon dioxide, methane, and many others—are on the rise, driving atmospheric warming, and on track to warm the atmosphere and oceans even more over the coming decades. This is in large part due to our dependence on fossil fuels and a variety of industrial processes. We're taking the greenhouse gases that had been stored in sediments for tens of millions of years and releasing them into the atmosphere in just a few centuries.

The physics of greenhouse gases prompts many people, especially scientists, to see a single, simple, solution: drastically reduce CO_2 emissions. What could be more obvious? But it turns out that in recent centuries we've woven our dependence on fossil fuels tightly through everything we do, starting with food production and water use, and we've invested trillions of today's dollars in energy infrastructure tailored to continuing that dependence. We can't take that step in isolation. Hitting the CO_2 off-switch is inherently a social problem. Viewed through that lens, it shows an entirely different look and feel.

Fifteen years ago, social scientists were asked to provide just such a perspective on climate change. The result was a four-volume assessment of the social science research relevant to global climate change that's been in print for more than a decade. Titled *Human Choice and Climate Change*, edited by Steve Rayner and Elizabeth L. Malone, this truly extraordinary effort centered on a Vancouver meeting in 1997 and drew on more than one hundred contributors. Especially intriguing was a small satellite document issued with the assessment titled "Ten suggestions for policymakers."

Rayner and Malone started this way:

What can public and private decisionmakers learn from a wide-ranging look at the social sciences and the issue of human choice and climate change that illuminates the evaluation of policy goals, implementation strategies, and choices about paths forward? At present, proposed policies are heavily focused on the development and implementation of intergovernmental agreements on immediate emissions reductions. In the spirit of cognitive and analytic pluralism that has guided the creation of *Human Choice and Climate Change*, we look beyond the present policy priorities to see if there are adjustments, or even wholesale changes, to the present course that could be made on the basis of a social science perspective. To this end we offer ten suggestions to complement and challenge existing approaches to public and private sector decision-making:

1. View the issue of climate change holistically, not just as the problem of emissions reductions.
2. Recognize that, for climate policymaking, institutional limits to global sustainability are at least as important as environmental limits.
3. Prepare for the likelihood that social, economic, and technological change will be more rapid and have greater direct impacts on human populations than climate change.
4. Recognize the limits of rational planning.
5. Employ the full range of analytical perspectives and decision aids from natural and social sciences and the humanities in climate change policymaking.
6. Design policy instruments for real world conditions rather than try to make the world conform to a particular policy model.
7. Incorporate climate change into other more immediate issues, such as employment, defense, economic development, and public health.

8. Take a regional and local approach to climate policymaking and implementation.
9. Direct resources into identifying vulnerability and promoting resilience, especially where the impacts will be largest.
10. Use a pluralistic approach to decision-making.

Rayner and Malone then expanded on these statements in a tightly crafted 40-page document. Fifteen years later, it is worthwhile to take stock and reassess these statements and supporting analysis. How well have they stood the test of time? Do they still seem fresh? Or do they feel dated? What has been their impact on the policy process to date? What about going forward? Do they offer intriguing starting points for policy formulation? Could they guide or constrain the search for viable policy alternatives? How has the status of social science research changed in the years since?

The short answer: These suggestions looked to be on point when they were issued, and they continue to look spot on today. Are they widely used? That may be a different story. Perhaps they're used in a fragmented way, but not so much as a framework. And attribution is rare. But here's one small exception, going back to the Stern Review on the Economics of Climate Change[3] published in 2006.

When the Stern Review came out, Professor Mike Hulme, director of the Tyndall Center and School of Environmental Sciences of the University of East Anglia, prepared a comment for the British Ecological Society *Bulletin*. Using the Rayner–Malone assessment as a yardstick, Hulme predicted that the Stern Review would have little impact. He asked:

> But how effective will it [the Stern Review] be and what difference will it make? In ten years time, will we be able to look back and analyse a pre-Stern and post-Stern discourse about climate change, or see the 2006 marking some break-point in climate policy?

---

3. The Stern Review was written by Nicholas Stern, an economist and chair of the Grantham Research Institute for Climate Change and the Environment at the London School of Economics. It assessed, for the British government, the anticipated future effect of climate change on the world economy. It concluded that failure to reduce carbon emissions (business as usual) would adversely affect food production, water resources, public health, and the environment that in turn would lead to future annual GDP losses of 5 percent or more. Mr. Stern urged investments something like 1 percent a year of GDP (which he increased in 2008 to 2 percent) to forestall this outcome.

I suspect not. To look for the reasons one need do no more than re-wind the clock to 1998 and the publication of the proceedings of the largest co-ordinated exercise yet undertaken by social scientists into examining the implications of climate change for human choice (Rayner and Malone, 1998). A self-proclaimed 'complement' to the United Nation's IPCC, this five year assessment delivered ten suggestions for policymakers in regard to climate change. *They deserve wider visibility and recognition* [emphasis added]. To understand the limits of The Stern Review let me mention just three of these ten suggestions, all of which emerged from an extended examination of knowledge emerging from the social sciences (and anthropogenic climate change after all has emerged from society, not from nature):

- "Recognise that for climate policy-making institutional limits to global sustainability are at least as important as environmental limits." The Stern Review has very little to say about new institutional arrangements commensurate with the nature of climate change decision-making. The barriers to effective action on climate change is not incomplete science or uncertain analysis, but the inertia of collective decision-making across unaligned or even orthogonal institutions.
- "Employ the full range of analytic perspectives and decision aids form the natural and social sciences and humanities in climate change policy-making." The Stern Review remains dominated by natural science and macro-economic perspectives on decision-making and although some concession to the role of values and ethics is made in the review, the values and ethical judgements made are pronounced rather than negotiated.
- "Direct resources to identifying vulnerability and promoting resilience, especially where the impacts [of climate change] will be largest." The Stern Review continues to place emphasis on linear goal-setting and implementation; a more strategic approach is to focus on measures that promote societal resilience and opportunities for strategic switching, informed by regional and local perspectives.

The tragedy is that social scientists have a lot to offer a world seeking to make full use of Earth observations, science, and services in order to achieve safety in the face of hazards and sustainable development and natural resource use. If more robustly underwritten, social scientists could contribute far more. But what they have offered has all too often gone ignored, and prospects for substantial budget increases for social science look slim indeed.

3.2. Social *Change*

Once in the late 1970s when I was flying to Atlanta, I found myself sitting next to a gentleman who spoke (by my lights) with an accent. "Where are you from?" I asked. "The Union of South Africa," he answered.

"What are you doing here in the United States?"

"Looking for a job."

"Why here?"

"Because my business is long-range strategic planning for multibillion-dollar corporations,"[4] he said. "For my work the starting assumption is that the host society is stable. In the Union of South Africa [still under apartheid at that time] that is such an unrealistic assumption that my job is without meaning."

Hmm. We talked much longer—and about much more—but that was the piece of the conversation that made the biggest impression on me. I've had occasion to reflect on it many times over the years. And here is the conclusion that returns each time: In the world of the 21st century no nation is stable in my companion's sense. In this rapidly evolving world, long-range strategic planning has either morphed into something else or it is of no value.

Here is a brief recap of the social change currently underway worldwide (Chapter 4 provides some slight additional perspective). Human populations have increased rapidly over the past 100 years. At the same time, those populations have concentrated in cities—not just cities but megacities. The world's population not too many generations ago was rural; today, more than 50 percent of the population is urban. Two social changes have driven and accompanied this migration. The first is industrialization, economic globalization, and corresponding job growth. The second is the development and widespread dependence on critical infrastructure: electricity and telecommunications; natural-gas pipelines; water lines and sewage systems; and roadways (for horizontal transportation) and elevators (for vertical transportation). With economic development, birth rates are declining. Populations in the developed world are aging rapidly.

At the same time, the large majority of the world's peoples now live in manmade settings, one or two steps removed from real-world conditions. This merits a deeper look.

---

4. In my 1970s naiveté I was surprised to learn that the Union of South Africa had any corporations that large, let alone several.

3.3 The Age of Virtual Reality[5]

Here's a story we all know full well, a story told and retold in print and on the artist's canvas— the moment when Adam and Eve were unceremoniously evicted from the Garden of Eden:

> The Lord God made garments of skin for Adam and his wife and clothed them. And the Lord God said, 'The man has now become like one of us, knowing good and evil. He must not be allowed to reach out his hand and take also from the tree of life and eat, and live forever.' So the Lord God banished him from the Garden of Eden to work the ground from which he had been taken. After he drove the man out, he placed on the east side of the Garden of Eden cherubim and a flaming sword flashing back and forth to guard the way to the tree of life. —Genesis 3:21–24 (NIV)

You and I know only too well that we live on the *real* world—not in any Garden of Eden. In the Garden of Eden, resources were both limitless and free. The environment was unchanging and benign, as were all of its creatures. There may have been labor, but it was labor without vexation and it never lacked purpose. It bore fruit. By contrast, in this real world, labor can bring thorns and thistles, dissatisfaction, and no end of weariness. We can feel the boss doesn't value our work, and the world doesn't value his or hers. Our company or our federal agency or our university might be dysfunctional and might struggle to keep up with the competition. We can be overwhelmed by the brokenness we see around us.

In this real world, extraction of resources (water, food, energy, materials) comes at a price—destruction of habitat, landscape, the environment, ecosystems, and biodiversity.

There *is* much good and much to like, but for most of us, most of the time, the curse is that we're blinded to the good side and see with way too much clarity the bad. Doomsaying is a growth industry.

In this real world, extraction of resources (water, food, energy, materials) comes at a price—destruction of habitat, landscape, the environment, ecosystems, and biodiversity. And this real world never proves entirely benign.

---

5. The second in notional names for the age we live in. Excerpted and adapted from a post on LivingontheRealWorld, dated April 20, 2011.

Conditions are rarely just right. Most of the time, we're suffering drought or flood, extreme cold or scorching heat. Storm clouds always threaten on the horizon. At any instant the earth beneath us can shake us to death, swallow us up, trigger a tsunami, let loose a mountain of magma with violent fury. The real world does its business through extremes—and most of these extremes pose a danger.

The Garden of Eden? The image is always near at hand. We see it in our mind's eye. We can't get it out of our head that sometime, someplace, there was a better existence, and somehow we either lost it or were kicked out. Metaphorically, we keep trying to break past those cherubim to muscle our way back in.

Or, alternatively, we attempt to construct our own, new Garden.

For present purposes, it helps to think of three phases of this Back-to-the-Garden-of-Eden Project in human history. The first Back-to-Eden Version 1.0 includes the cultivation of crops and domestication of livestock—and close behind, the birth of civilization, some 10,000 years back. According to archaeologists, anthropologists, and evolutionary biologists, it took us maybe 100,000 or 1,000,000 years to get to that point, but we did it. We got past the clans and the tribalism, and we constructed towns, and then cities, and then nations, and then empires.

Version 2.0: Maybe 10,000 years later—over the past 200 to 300 years or so—the Industrial Revolution. As I write this, and probably as you read it, we're in a controlled environment, a virtual climate that we've engineered to our liking. I'm writing in a heated/air-conditioned office, looking out at an urbanized, high-rise landscape. People are walking from place to place but only for the shortest of distances. A steady stream of cars and buses carry the street-level traffic. Ten or 12 stories down, the Metro rumbles past every few minutes. The water comes from a tap and is always deemed safe to drink. Food is plentifully available, representing a world of cuisines from myriad vendors and restaurants. Surely this is Eden.

What about Version 3.0? We've been in a third-level reality this past decade or so, and the pace of innovation and social change is gathering speed. Remember those people on the street in Eden? They're talking, but look and listen closely. They're not so much talking with each other; they're on their smartphones, reaching out to friends and relatives who may be across the city, or even half a world away. And though I have an office window, it's not entirely an asset. I have to keep focused on the LCD monitor before me. In fact, I'm better off closing the blinds and shielding my computer monitor from the outside glare. *Why do I even need the outside view when*

I can call up the world's most beautiful scenery in high-definition, even in motion, with a few key clicks? And more and better technology is coming online. Smarter phones. Streaming video. More capable and extensive social networking. Shopping. Games and recreation. Entertainment. Telecommuting. The world's information, the world's pulse— everything I need—at my fingertips. It doesn't get any better than this.

Not so fast. Even as the developed world is settling in to this human-constructed Eden, Version 3.0, old troubles stubbornly hang around, even as new signs of emerging trouble intrude via those same smartphones and laptops. We're not getting along with the person on the other end of that phone any better than we did when we met face-to-face. That physical fitness we maintained pre-Eden just trying to stay alive? Now we have to go to the gym. Work? It's still aggravating and broken. We can be tasked, even without our permission, in multiple ways. We can still expend massive amounts of effort only to see our toil come to naught. Here in the United States, while we were struggling to make a go of it at our workplace, a handful of Wall Street wizards found a way to use the Internet to construct a whole bunch of financial instruments that somehow lost the rest of us 300 million people trillions upon trillions of dollars in just a few months. Just a sample, but you get the idea.

The biggest problem, though, is that each version of a new Eden has added an additional degree of separation between us and the real world on which we're living. The Internet information stream is hinting that the real world on which we depend is not so robust in the face of seven billion of us tromping about as we might have hoped. After years of ample stores of food supply courtesy of the green revolution, spot shortages and rising prices are resurfacing. Groundwater supplies are being depleted. Our city water supply has been cleaned up from earlier use upstream. Chemical analysis reveals it to be loaded with endocrine disruptors, antibiotics, and more. Habitat is disappearing, and with it many endangered species. Resource extraction is getting more problematic (think the Chilean mine disaster or the BP oil spill or the Fukushima reactor collapse). Natural disasters and their consequences seem to be getting bigger in scale and impact. But for the majority of us, the information has been sanitized into bits and bytes, a mere abstraction. We struggle to comprehend the enormity of what is going on. Only a handful of us are experiencing these problems on the ground. Many of those best positioned to see real-world problems firsthand are in such straitened circumstances—malnourished and/or living in poverty—that they are in no condition to care or sound an alarm. This real world has been relegated to a place we mostly visit as tourists, rather than a place where we work. All this is occurring while we seem to be losing our ability to trust, to be committed, to love one another, which is the basis for any kind of chance we have to work together and meet these challenges.

To review: We are in the midst of profound social change:

- We're transitioning from a society where most people live in daily, direct contact with the land to one where that contact is indirect and removed not by one but by two degrees of separation.
- We're making this transition at precisely the point in history when human impact on the skin of the planet has greatly accelerated—an Anthropocene where our choice is not whether to manage the planet but whether to manage it well.
- We're making this transition at the same time that social factors impact conditions at a distance, where the choices available to the African farmer are dictated not by local weather conditions and changes in those but by market and technological forces half a world away and where the concentration of wealth is not limited to local exchange but is instead truly global.
- We're making this transition at a time when our circle of interdependence extends further than our circle of awareness and *far* further than our circle of trust.
- We're making this transition at such rapid speed that no one has had the opportunity to internalize the new realities.
- We're making this transition at a time when centralized, top-down command and control approaches no longer work and when nothing has yet replaced them.

If all this sounds wickedly complicated, that's because it is.

3.4. Wicked Problems

You and I will succeed in life only to the extent we are socially realistic. However, the fact is, most of our social problems—at least the ones that matter—are what social scientists call *wicked*.

And not the slang use of *wicked* often heard in casual conversation used for mere emphasis. But the true definition of wicked, the kind to classify, say, the task of simultaneously solving the so-called Navier–Stokes equations, the set of coupled equations that are at the heart of weather prediction. Weather prediction? Surely that's a wicked problem! Ah, but now the astrophysicists scoff. The stars they study consist of plasma—a gas of electrically charged ions and electrons. To model stellar behavior requires taking account of electrical and magnetic forces, and incorporating additional equations—making a true jumble of equations to be solved simultaneously. What could be harder?

But here's the reality. The hardest problems—the truly wicked, the most intractable—come not from the "hard" but from the "soft" sciences, those that deal with how our brains work and the way large numbers of us engage each other. Sociologists use this term to describe a class of challenges that societies find themselves poorly equipped to overcome.

Horst Rittel formulated this notion in work dating to 1967, but for a more readable account, and one that makes application to environmental problems, take a look at Steve Rayner's 2006 Jack Beale lecture. Rayner listed six attributes (Rittel had a rather larger number) that make a social problem "wicked":

- *Characteristics of deeper problems.* Mr. Rayner gives a great example here. How to explain educational underperformance? Well, poverty plays a role. The few hours of each day children spend in school doesn't compensate for home environments that are abusive, poverty-stricken, drug-ridden, and dysfunctional in other ways. What's the root cause of poverty? A caste system contributes. And what factors determine caste? Education, for one. See? The arguments tend to get circular.
- *Little opportunity for trial-and-error learning.* One example is mitigation approaches to climate change. Make Draconian reductions in CO_2 emissions and *maybe* 100 years from now we'll be glad we did. Well, maybe. But we can't run a controlled experiment to make sure.
- *No clear set of alternative solutions.* Our problem is not choosing among, say, three policy options: should we build the superhighway here? Or there? Should we choose wind energy instead of nuclear power or fossil fuels or simply conserve? No one approach to energy policy seems to get us where we need to go.
- *Contradictory certitudes.* We hear a lot these days about how scientists need to treat and communicate uncertainty. But when it comes to wicked problems, curiously enough, most people aren't in doubt. They *know* how to eradicate poverty. Squash terrorism. Cope with climate change. Take that last one. The trouble is former Vice President Al Gore and Senator James Inhofe—picking two names totally at random—don't agree. In fact, seven billion of us will give you seven billion opinions.

- *Redistributive implications for entrenched interests.* You and I may say it's important in principle to pay the true cost of burning fossil fuels, but in practice we're loathe to pay that cost to either a utility company or the government. We'd rather hold onto the cash for ourselves, even if for just a bit longer.
- *No solutions, just coping strategies.* This is akin to the problems faced by diabetics in that there's no cure, only therapies. (And, significantly for our narrative, as we'll see later on . . . very similar to the problem the Navier–Stokes equations pose to meteorologists.)

By some lights, Mr. Rayner's list of characteristics, daunting though it is, stops short of the true difficulties here. Others have since noted some additional aspects that make challenges such as climate change even more problematic. These so-called "super-wicked" problems include attributes such as the following:

- Time is running out.
- There is no central authority.
- Those seeking to solve the problem are causing it.
- Policies discount the future irrationally.

Here's a partial list of some of the global challenges that arguably fall into this "wicked" category: preservation of biodiversity; stewardship of marine resources; deforestation; climate change; disease; poverty; terrorism; war; education; natural hazards; tax codes; and intellectual property law.[6] You might want to relabel or combine some of these. You might want to add a few of your own.

Note that Mr. Rittel and Mr. Rayner, and their fellow social scientists, by compiling such a list of characteristics haven't really offered a solution to any of these real-world problems. But they have performed a valuable service. They've sharpened our thinking. They've forced us to confront just how fundamental are the issues that make these problems thorny. They motivate us to cope with these big problems by cracking one or more of the barriers at

---

6. If this list looks partial, that's because it is. These are but a few of the problems discussed by J. F. Rischard, in his 2003 book *High Noon: Twenty Global Problems. Twenty Years to Solve Them.* This is perhaps the best book ever written you've never heard of.

a time.[7] Rayner, in his Jack Beale lecture, has pointed out that one attribute looks vulnerable: the lack of opportunities for trial-and-error learning. He suggests that framing the climate-change issue in terms of adaptation rather than mitigation alone breaks the one global problem into myriad smaller ones. We'll return to this notion in Chapter 9.

3.5. Spiritual Realities

Let's look at the attributes that make a social problem wicked through a different lens. Let's ask which of these characteristics might prove more tractable, or be less of an obstacle to good societal outcomes, if we were somehow *better*.

That's not better as in smarter. That's better as in better-natured, more virtuous, more caring, more loving. How about putting the welfare of others ahead of our own. Being more honest. Being more faithful at doing unto others as we would have them do unto us. Being more fair, honorable, and responsible.

At a minimum, attributes making a problem wicked would include *contradictory certitudes* (which arises from our inflated idea of our own mental capacities), *redistributive implications for entrenched interests* (our tendency to see our own interests as in competition with others, and preferentially seeking our own ends), and (from the expanded list) *those seeking to solve the problem are causing it*.

In many circles, these problems would be viewed as a portal to spiritual realities, a third set of realities as distinct from social realities as social realities are from the physical. In other quarters, to give spiritual realities any separate, independent standing is anathema. We see something like this state of affairs at the intersection of physics and astronomy. To make the astronomical observations work, physicists have inferred the existence of dark matter and dark energy. According to cosmologists, dark matter neither emits nor absorbs light or any kind of electromagnetic radiation. It nevertheless affects matter (as we know it), electromagnetic waves, and the structure of the universe. And it doesn't just exist in trace amounts. Physicists tell us that to make the sums work, the total mass-energy (remember, Einstein

---

7. The approach is similar to that adopted by cancer researchers. Though despairing of a cure, they focus on buying time, on improving methods for delivering toxic chemicals to the tumor site rather than dispersing them throughout the patient's entire body, on targeting specific weaknesses of cancerous cells, etc.

quantified the equivalence of mass and energy) of the universe has to be apportioned roughly this way: 5 percent ordinary matter (the stuff most of us learned about in school), 27 percent dark matter, and 68 percent dark energy. Dark energy is a hypothetical form of energy that permeates the universe and tends to accelerate its expansion.

It may be that 100 years from now, we'll still be using this terminology. But it's also possible that by then we'll consider dark matter and dark energy to be something more distinct from their ordinary counterparts. In the same spirit, you may prefer to see what some call spiritual matters as an extended form of social reality. Or you may prefer to see them as entirely separate. But by whatever name, we'll be circling back to them.

3.6. Recap

> Life is like an onion: you peel it off one layer at a time, and sometimes you weep. —*Carl Sandburg.*

Let's see where we stand. Successfully living on the real world requires that we master and accommodate three realities, simultaneously:[8] physical, social, and spiritual.

The physical reality is that seven billion of us are essentially confined to Earth's surface until further notice. We'd like it best if we could provide the lifestyle enjoyed by the most well-off to all seven billion and if we could all enjoy that lifestyle indefinitely. But the best we can do is buy time. And the fact is that we're dangerously ignorant. We don't know how much time we have. It's urgent that we get a better handle on that, innovate our way to more breathing room, and accomplish this quickly.

---

8. What does *simultaneously* mean? Here's an analogy. In crossword puzzles, the correct answers are only those words that simultaneously fit both the across and down definitions. It's not enough that you have a five-letter word for "barnyard animal" (goat or pig or duck or lamb or cow clearly don't work, but chick or goose or horse or sheep or piggy or even mouse might all fit); the third letter also has, say, to be the fifth letter in a six-letter word denoting "former explosive device." That might be petard, in which case "horse" might be the correct answer. To gain more appreciation for what meteorologists are up against, now picture a five-dimensional crossword puzzle, one that's not confined to a newspaper page but is the size of Earth and filled with clues to words that must all be solved every few minutes and then solved anew, with fresh definitions and patterns. We'll take another look at this definition in Chapter 7 and again in Chapter 9.

The social reality is that this goal can't be achieved by addressing physical science and engineering alone. People are involved, all of us bringing to the table a broad spectrum of ideals, encompassing both self-interest at one extreme and a sense of fairness at the other. Living on the real world therefore demands a framework or multiple frameworks at local, national, and global levels that make our self-interest congruent with our noblest values.

The spiritual reality is that we have difficulty sublimating our perceived self-interest to a larger common good unaided, on our own. This merits a chapter of its own, but that is for another time, and perhaps another book.

Now let's peel off another layer of the onion. Living on the real world requires that, subject to physical and social (and spiritual) realities, we meet three challenges *simultaneously*: gain the energy, water, food, and other resources we require to live for today; protect the environment and ecosystem services we need to maintain those resources indefinitely; and build resilience locally and globally needed to weather nature's extremes. Again, the evidence we've accumulated suggests it's far easier to achieve any two of these goals simultaneously at the expense of the third.

Cue the weeping?

That would not just be premature but uncalled for. Meteorologists (among others) have developed approaches to a similarly intractable problem: solving *simultaneously* the so-called Navier–Stokes equations that govern atmospheric dynamics. More precisely, meteorologists have not solved these differential equations. They can't be solved in closed form. But meteorologists have approximated them with arithmetic equations that can be solved piecemeal. This sounds similar to the coping strategies Mr. Rayner referred to in his sixth characteristic of wicked problems. Maybe meteorologists are onto something. We'll see later how this approach might be applied to the larger challenges we face.

But for now, let's look a little more closely at those challenges. Let's start by constructing a persistence forecast for the next 100 years. Let's address the question: What kind of future is likely if we take no action?

4

A PERSISTENCE FORECAST FOR THE 21ST CENTURY

Years ago, when I was still at the National Oceanic and Atmospheric Administration (NOAA), I worked for Joseph O. Fletcher.

Joe was a truly extraordinary man. He had flown B-24s for (what is now) the U.S. Air Force in World War II and the years immediately following. He established a weather station on an Arctic tabular iceberg in 1948, and he was part of a team that landed a plane on the North Pole in 1952. After leaving the Air Force, he served for a spell as director of NSF's Office of Polar Programs and then joined NOAA.

Joe loved the study of the Arctic and climate. Say "good morning!" to Joe, and he'd lean back and say, "*Speaking* of climate . . ." and then launch into discourse. Close behind, his favorite activity was asking questions to test his subordinates. For a while, back in the 1980s, I was one of his preferred targets.

One morning, he asked, "Bill, how do you know that in 10 years we're not going to have a fleet of Supersonic transports (SSTs), with most commercial jet travel occurring at supersonic speeds?" (There had been some buzz about this at the time; Joe favored questions on the contrarian side.) I came up empty and grumbled something inadequate, to which he replied, "Because none of the things needed to lead to that is underway now."

By that he meant that the designs for the planes weren't on the drawing boards. The orders for the planes weren't in the hands of the airframe manufacturers. The only facilities that could have produced such planes were committed for years to manufacturing more conventional aircraft. The needed runways weren't under construction and so forth. A global proliferation of SSTs was therefore not in prospect.

Joe was saying, "In this case, persistence is the best forecast."

4.1. The Human Race Is on a Roll

At temperate latitudes, if you want a weather forecast that is accurate about two-thirds of the time, simply predict that tomorrow's weather will be the same as today's. Joe Fletcher's real-world outlook was founded on a similar idea. In other settings, this is sometimes referred to as the business-as-usual scenario. By looking back a bit, and being realistic about our present circumstances and the trends responsible, we can draw inferences about the future—especially about the kind of future that's likely if we take no action—if we stay on our present course.

How does business-as-usual look for the world as a whole? Pretty good, actually, at least superficially.

You could say, and you'd be perfectly right, that our present world and lives are problematic. We're anxious and driven. There's no peace in the world, either for peoples or for individuals. There's illness and suffering and death. There's pollution and urban sprawl and blight in the inner cities. There's poverty and terrorism and outright war. As discussed previously, we tend toward such pessimistic views.

But please put these considerations aside for the moment. Instead, let's look at more material metrics.

Here are three. In a short period of time, we humans have:

1. greatly increased our numbers;
2. increased our per capita resource consumption; and
3. accelerated the advance of science and technology and the pace of social change.

By these three criteria, the past two centuries have been a rousing success.

Figure 4.1 shows a graph of human population through thousands of years of our history. Over the past 100 years or so, our numbers have really taken off.

If you were to look at a graph of water consumption per capita in the developed world only, you'd find that the graph had about the same shape over about the same time period as the population graph. It would be going along pretty much level at a low value for tens of thousands of years, look-

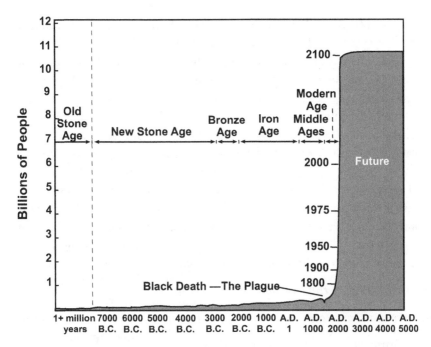

FIGURE 4.1. World population growth through history, courtesy of Population Reference Bureau.

ing constant. Then, over the last 100 years or so, it would spike sharply upward. Today, you and I use 400 times as much water as our subsistence-level counterpart (nearly 200 gallons per day versus 2 quarts). To emphasize, this increased per capita consumption is a relatively recent phenomenon, going back only 100 years or so.

Our physiological requirement hasn't changed. It remains about that same couple of quarts of water per person per day needed by our subsistence-level counterpart. The exact figure depends on body weight and the level of physical activity, but that's about it. But today we use water for so much more. We wash dishes and laundry. We use flush toilets. If we have a lawn, we water it, to keep the grass and all those shrubs and decorative plants looking good. We eat meat, and it takes a lot of water to raise the livestock we annually devour. We use electricity, and it takes prodigious amounts of water to generate that power. We drive cars, and it takes a lot of water to manufacture that car.

It's not just water that we slurp up. If you look at our consumption of power per capita over human history, that graph would also appear about the same. In Benin and Cameroon, people use 300 to 600 kilograms-of-oil-equivalent (kgoe) per year (a pound or so of oil per day). Here in the United States, we use 8,000 kgoe per year (20 times as much).

About the same is true for steel. Over the past 100 years, steel production worldwide has increased from about 30 million tons to about 800 million tons—a factor of 20 or so. Population hasn't increased by anything like that amount, so the per capita consumption, especially in the developed world, has skyrocketed. Substitute any natural resource—iron, nickel, lead, copper, platinum, selenium, or molybdenum—and the result is qualitatively similar.

Get the idea? If we're trying to look at the footprint of the human race on Earth's surface, it's not just taking into consideration that the population went up by a factor of 4 over the past 100 years. With the rate of consumption of almost everything on the rise, the virtual population rise (how many people living the old way it would take to consume resources—water, energy, raw materials—as fast as we're now consuming), it's as if the old-style, pre-industrial population

Soon the poorest billion of the world's population, still living at the old, subsistence levels of resource consumption, will start using resources at the same rate as the wealthiest of us.

had multiplied by a factor of 50 to 100. And higher rates of consumption still may be in the wings. Soon the poorest billion of the world's population, still living at the old, subsistence levels of resource consumption, will start using resources at the same rate as the wealthiest of us.

Things get even more interesting when we look at the pace of innovation and social change.

Take mechanical invention. It's hard to put a precise date on the invention of the wheel, but let's say it occurred approximately 10,000 years ago. That means, depending upon where you put the exact origin of the human race, it took quite a while to invent the wheel: somewhere between 200,000 and 1,000,000 years. But, after the wheel was invented, it only took 10,000 years to invent the car and the airplane. Less than 70 years after inventing the airplane, we'd built rockets and gone to the moon. The pace is picking up and by quite a bit.

The same is true for medicine. Around A.D. 130–150, the great physician Galen announced that the function of the heart was to heat the body. He thought of it as a furnace. You can appreciate the reasoning. The heart stops beating, and our temperature drops from 98.6° Fahrenheit to room temperature. We cool off.

It took 200,000 years to get to Galen's insight. It wasn't until the 16th century, 1,400 years later, that William Harvey suggested otherwise: The heart is not a furnace but a pump. And 50 years ago, more or less, Watson and Crick announced that the molecular structure of DNA is a double helix. Today, only half a century later, people are cloning sheep, making pets that glow in the dark, genetically modifying food, testing DNA in high school classes, and using DNA to predict our susceptibility to disease and to identify and prosecute criminals. The pace of innovation is accelerating.

The ramp-up of computation is particularly dramatic. Again, it essentially took 200,000 years to invent the counting board and then the abacus. We start seeing examples of these only a few hundred years before the time of Jesus. Another 2,000 years were needed to develop the first computers able to perform complicated calculations, such as global numerical weather prediction. These didn't appear until around 1950. They each used some 10,000 to 20,000 vacuum tubes and filled large rooms. In the last half-century, however, we've progressed to PCs, iPhones, and iPads. Your cell phone is many, many times more powerful than these first-generation computers. The chip you find in a musical greeting card, which plays a tune when you open it up, is the equivalent to the room-sized computer of the mid-20th century. The most powerful computers of 50 years ago are today's throwaways.

So, innovation is like popping corn. It seems to take forever for that first kernel to pop. The second one doesn't take quite so long. And soon, the noise is deafening. That's the real world. The world we live in.[1]

Exciting.

It's not just the physical, chemical, or biological sciences that are progressing at such a clip. Social change is more rapid as well. For most of human experience, we were hunter-gatherers. There was very little specialization within the group, except perhaps between the men and the women. Everyone in a given family or clan was doing pretty much the same thing. The switch to agriculture 10,000 years ago—putting down roots and relying on grains we grew and animals we domesticated—accommodated and indeed fostered specialization. Most people farmed, but some could now make a living by crafting tools for farmers, throwing pottery for cooking, even crafting ornamental jewelry and the like. Towns and cities, and then civilizations, began to appear. With this change came a jump in our interdependence. We began to need and count on each other in different ways. But even as late as a few

---

1. For an interesting read in a similar vein, one which attributes the innovation to economics, see Matt Ridley's *The Rational Optimist*.

centuries ago, things hadn't changed that much. When this country was founded, more than 90 percent of the 3 million population lived and worked on a farm. Today that statistic is more than reversed. Farmers make up less than 3 percent of the workforce. Technological changes, such as big farm machinery, mills, factories, railway transportation, and still other advances, moved people off the farm and packed them into towns and cities. It took virtually all of our 200,000 years to invent the elevator, but it made high-rise building practicable. Today more than half the world lives in cities, and their high-rise profiles are major features of the landscape.

We shouldn't gloss over the implications of rapid social change. It has not only accompanied but enabled our current population level as reflected in that earlier population graph. At seven billion strong, the world's peoples, if spread evenly on the earth's land surface area, would each occupy a plot something like 500 feet by 500 feet square—maybe 6 acres.

At seven billion strong, the world's peoples, if spread evenly on the earth's land surface area, would each occupy a plot something like 500 feet by 500 feet square—maybe 6 acres.

Picture that distribution. Would any of us find ourselves lucky enough to be on a plot that would support us, affording us a measure of self-sufficiency? Allow us to grow all our own food and raise livestock as well? Yield materials for building shelter? Meet our energy needs?

No way. It's hard to imagine anything sustainable on this basis.

The secret to accomplishing a seven-billion level of population has been teamwork and trust, a massive shift from independence to *interdependence*.[2] In just a century or so, we have individually and collectively made a transition from going it alone to counting on each other. And we count on each other not just from time to time, say, on special occasions or once a week or so.

We count on each other moment by moment. For example, as I type an early draft of this chapter, I'm sitting in a hotel room in San Francisco during the predawn hours before the conference I'm attending gets underway for the day. Without hotel staff making sure the building is heated I couldn't be working. And that hotel staff is in fact depending on natural gas supplied by countless companies stretching across the globe. There may be a technician

---

2. As indicated in Chapter 1, one name for the age we live in.

or two in the building who is at the ready to deal with any problems that might arise, but chances are they're not actively doing anything at this particular instant. I'm dependent on electricity in the same way. The instant the electricity goes off is the instant I have to shift my entire focus from this book to groping my way down darkened emergency staircases in the building.

In an hour I'll hit the streets to find a Starbucks and breakfast. Hundreds of other coffee houses and restaurants are competing for my business. A disruption of gas, or water, or electricity brings that to an abrupt halt.

An earthquake would do the trick. A repeat of the San Francisco earthquake of 1906 would kill many of us and put the rest on the streets in full survival mode. But it wouldn't of course take anything that dramatic. A few years ago, I was at this same conference when the meeting and many other downtown San Francisco activities came to a screeching halt. An electrical crew had inadvertently left a three-foot spanner lying across some lines when they powered back up after some routine maintenance. The resulting short circuit blew the substation transformers. Everything and everyone had to mark time in the dark, essentially confined to hotel rooms and office towers until power was restored. The commute home that evening was a nightmare.

Yesterday's nation of farmers represented essentially independent households. Any adverse weather of the day impacted multiple farms, but each farm impact was essentially an independent, localized event. By contrast, in the fall of 2012, when Hurricane Sandy flooded a portion of the New York subway system, it didn't just disrupt the few adjacent city blocks that were actually under water. It brought a halt to the smooth functioning of the entire metropolitan area, even though virtually the whole of the city was no more than windblown and damp.

Other social change is also underway. Look more closely at cities like New York. It used to be that most city dwellers worked in manufacturing. Today's societies are shifting to service-based industries, for example, the financial sector. Entertainment, travel, tourism, and recreation also make up a huge percentage of the economy. In the past, social activity had to be physical; today much social activity is virtual, accomplished at a distance electronically. Societies used to be populated by the young. Families were large, and many people died in childhood, or in childbirth for that matter. Most of the increase in life span has been accomplished by reducing death in childhood. And with birth control and improved educational and career opportunities for women, a sharp drop in family size has come to pass. That has contributed to another demographic shift. Developed societies are now older on average and are aging rapidly.

Another major change is that businesses used to be local. Today they're national and global. Businesses used to maintain large inventories; today they use just-in-time manufacturing methods.

To repeat: In a short period of time, the human race has greatly grown in total numbers; has radically increased its per capita use of resources of every type; and has accelerated the rate of social change and scientific and technological advance.

4.2. What Qualifies as a *Short* Period of Time?

Human success has been achieved in a *short* time by five comparative measures.

1. *To start, we're talking about a short period of time compared with the geologic climate variability.* In the climate-change debate, much has been made of this. Earth's climate has at times over the past few billion years been either much colder or substantially warmer than it is today. But for the last 10,000 years, the climate has been fairly stable.

And it's only over the past 200 years that people have taken a serious intellectual interest in the whole notion of *climate*, which some define as *the composite or generally prevailing weather conditions of a region, as temperature, air pressure, humidity, precipitation, sunshine, cloudiness, and winds, throughout the year, averaged over a series of years.* This means that we have had time to tune our (relatively short-lived) success in agriculture, water resource management, and the like to only a rather narrow range of climatic variables. (For that matter, it means that we've had time to tune our vocabulary and our accompanying thought processes to a notion that wouldn't have been so useful during other, more variable periods; we'll return to this point later.)

2. *We're talking about a short period of time compared with the recurrence (return time) of extreme events.* For example, the New Madrid fault, centered in Missouri, experienced three major earthquakes in 1812. Each of those earthquakes has been estimated to have exceeded 8.0 on the Richter scale, numbering them among the strongest earthquakes to have hit this country. When they occurred, only several hundred people were living in the area of the quakes. They saw the earthquakes(s) create a new lake, some 140 miles long. They even reported (possibly erroneously) that the Mississippi River briefly reversed course. These earthquakes in Missouri were so violent and their effects so far-reaching that they were said to have rung church bells in Boston.

Several million people now live in the general vicinity of the quakes, and many of the natural gas pipelines that service the northeastern United States run through this area. Today, the impact of such earthquakes would be many times greater. And the U.S. Geological Survey predicts a 10 percent chance of a similar earthquake hitting within the next 50 years.

That same region is also known as Tornado Alley. The biggest tornadoes rumbling through across the Great Plains cut a swath a mile or so wide that might tear up a stretch tens of miles in length. But the people in the Missouri–Illinois–Indiana area in 1805 reported that a tornado had knocked down trees, creating a barrier impenetrable to travel, over a stretch of wooded countryside 2 miles across by 200 miles long. Imagine the consequences of such a tornado today, running through St. Louis or Chicago.

Consider Mount Rainier. Every 10,000 years or so, Mount Rainier erupts. But that isn't the proximate danger. The more immediate threat is mudslides. The mountain is geothermally active, and the steam that courses through the mountain is acidic. It leaches the metal out of the rocks, weakening them, making the flanks of the cone prone to failure during earthquakes. Every few hundred years on average over the past 10,000, Rainier produces a mudslide: a slurry of rock, mud, ice, and water that races down the slopes at speeds approaching 100 miles per hour. Geologic evidence suggests that some of these slides have reached Puget Sound. However, since the last such event we've been busily developing the area. Today 100,000 homes have been built on the former mudslides. Each day, 250,000 people commute to work on roads that cross them.

The threat from hurricanes is changing as well. As recently as a century ago, hurricanes hitting the United States largely made landfall over rural areas. Today that's no longer the case. The coasts have been built up over the past several decades, during a time period widely acknowledged to have been a lull in hurricane activity. Today's hurricanes find the United States a target-rich environment. Prior to 2012, a hurricane storm surge of such magnitude that it would flood New York's subway system was recognized as an abstract possibility but considered a hypothetical threat. Hurricane Sandy made the hypothetical a reality.

Similarly, tornados pose a different hazard than they did a century ago. When I was new to meteorology, conventional wisdom held that tornadoes didn't pose a threat to urbanized, high-rise environments. The buildings supposedly interfered with the boundary-layer inflows widely seen as vital to the tornado's maintenance. Today, however, we have several examples of tornado damage to city centers, and we recognize that urbanization and city growth have occurred so rapidly across the continent that we are only now starting to see the vulnerability. That earlier conventional wisdom is beginning to look less like fact and more like an urban legend.

3. *We're talking about a short period of time compared with the time required for the emergence of unintended consequences.* Think about your daily experience reading the newspaper. You find articles about lead, arsenic, and a potpourri of hormones in drinking water; the drawdown of aquifers; the pollution of groundwater; acid rain; and climate change—environmental degradation in hundreds of forms, at thousands of locations, on local, regional, and global expanses. None of these events was a national goal. No one 50 years ago said "Let's eliminate honeybees," or "Let's pollute our groundwater," or "Let's turn our rivers into a mixing pot for pharmaceuticals from city sewage and fertilizers, herbicides, and pesticides from rural runoff." These and many other crises were the unforeseen consequences of earlier attempts to solve this or that problem.

Acid rain is a good example. Years ago, people noticed that there was a lot of local environmental damage in the vicinity of smelters, power plants, and other facilities that burned fuel and put it out in short smokestacks. People who knew a little engineering and a little atmospheric science realized that simply by building taller smokestacks, it would be possible to put the stack effluent—all that bad stuff—high enough in the atmosphere that it would dilute over a large region before falling to the ground. At those lower concentrations, any pollution would be negligible. For decades, this seemed to work well. But then scientists, foresters, and others noticed that many of the effects that had been damaging on the local scale were starting to appear regionally. They were subtle but detectable, and definite. Acid rain had become a national—indeed, international—problem.

4. *We're talking about a short period of time compared with the time required to see whether we can keep it up.* (Science speak for this is our friend from Chapter 2: "sustainable development.") Suppose you'd reached what you thought was the finish line of a sprint, and the judge said that was just the initial 100 yards of what was really a 13-mile half marathon. You would find you couldn't maintain your pace. Even the best athlete would be hopelessly winded, attempting to sprint more than a short distance. With respect to our economic development and our efforts to feed, clothe, and shelter seven billion people, right now we *think* we can do it, but we can't be sure. For example, arable land is disappearing worldwide, a consequence of population growth and urbanization. Productivity per remaining acre is increasing,

thanks to improved agricultural practice and genetic modification of grains, but this is a continuing race, and it's not altogether evident at any given time just where the next advance is coming from, or whether continued advance is even in the wings.

5. *Finally, we're talking about a short period of time compared with the time required for seven billion people to internalize the changes in our circumstances.* This latter aspect is particularly important, and it is occurring at two levels.

First, there is the individual level. For all of human experience, many hundreds of thousands of years prior to the present day, everyone died in pretty much the same world into which he or she had been born. Social change was slow. Science and technological advance was hard won and was occurring at a snail's pace. And one of the truisms about the elderly was that they had much wisdom to offer the younger generation. So when the elderly were in charge, it wasn't so much of a bad thing. They had seen it all.

Today, by contrast, none of us is dying in the same world into which we were born. I was born during World War II. There were no commercial jet planes. No one had gone to the moon. Electronic computers didn't exist in any form. Telephones had dials, and most people were on party lines. Tape recorders coexisted with wire recorders. Tape cartridges were a thing of the future. If you had a college education, you had acquired all the knowledge you would ever need. Today, going to the moon is something we *used* to do. Tape cartridges have been consigned to history's dustbin along with Betamax tapes and floppy disks. And college education is only the portal to a lifetime of learning needed to stay relevant. You and I spend a large fraction of our time trying to keep pace with social and technological change. Our devices and ways of doing business are replaced with something better long before they physically wear out.

(Readers may be familiar with the financial concept of present-discounted-value. It states that 10 dollars 10 years from now is worth less to you and me than 10 dollars in hand. It further states that when inflation is high, the future promise of 10 dollars is worth less still. Something analogous happens to the value of knowledge during a time of rapid scientific and technological advance. Knowledge acquired in the past is worth less and less in the present, the faster the rate of change.)

Second, even as individuals struggle with this, so does our society as a whole. For example, it turns out to be a mistake to characterize climate change as a slow onset problem. Remember what we found in Chapter 3? Climate change is *rapid* onset, occurring in a short time with the time required for seven billion people to see it as a real threat, let alone agree on what to do about it.

To review, our success as a human race has occurred in a very short time, compared with:

1. the natural time scale of climate variability;
2. the recurrence time for weather extremes and other natural hazards such as earthquakes;
3. the time required for the emergence of unintended consequences of our success;
4. the time required to be sure we can maintain our success; and
5. the time required to internalize our changed circumstances.

4.3. Implications for the 21st Century

These five realities have five corresponding implications for our future.

1. The climate is going to vary, for both natural reasons and because of human influence, and (this is important) *we are going to regard that variation, whatever form it takes, as adverse*. Why? Because we have *tuned* our climate-sensitive activities—human settlement and water consumption, agriculture, energy production and use, and many other aspects of our daily lives—to a narrow range of climate conditions. At any given location, it doesn't matter whether the change is in the direction of hotter or cooler or wetter or drier, no one is going to like it.

2. We are going to continually be surprised by the severity of natural extremes because these recur on geological time scales long compared with a relatively brief human record. Where did the 2004 Indonesian tsunami come from? Memory of the last one was only dimly discernible in the cultural tradition of the region. What about the 2008 Sichuan (Chinese) earthquake that killed all the schoolchildren and left five million people homeless? And how about the eruption of the Eyjafjallajökull volcano in Iceland that shut down European air travel for weeks, the great Tohoku earthquake and tsunami of March 11, 2011, or the 1930s Dust Bowl in the United States? Paleoclimatologists, and volcanologists and seismologists looking at the geologic record, have returned from their research to report that these historic events are by no means worst-case scenarios. Bigger earthquakes have occurred in the past. The K-T meteor, which hit the Yucatan Peninsula, triggered a tsunami that reached inland in the United States all the way to what is now St. Louis. A few hundred years ago in the southwest desert of the United States saw dry spells that put the Dust Bowl to shame. The Mount St. Helens eruption

that captivated the attention of Americans alive at the time ejected a cubic kilometer of material into the atmosphere. The Toba Indonesia volcanic eruption of just 74,000 years ago (an eye blink in the geological scheme of things) ejected 3,000 *times* as much material into the air. More events like these are in our future. We're pathetically unprepared for them.

3. Coping with the unintended consequences of past decisions and actions is going to chew up more of our time and get more urgent and demanding. The rate of population growth, the increase in per capita use of resources, and the introduction of untried science and technologies are giving rise to unintended consequences that are not only accelerating but also interacting more strongly with each other. Take fracking, for example. This new technology for releasing natural gas from deep shales far below Earth's surface looks able to provide abundant, inexpensive natural gas that could meet our energy needs for more than a century. It promises to transform the United States from an energy importer to an exporter by 2020 and to energy self-sufficiency by 2035 or so. The industry claims that this technology does not pollute underground aquifers or trigger seismic activity. But evidence to the contrary is accumulating: Studies find contamination of groundwater, degradation of air quality, and the migration of gases and hydraulic fracking chemicals to the surface. By the time we fully comprehend all the negative side effects, the practice will be so pervasive both in the United States and worldwide that dealing with the consequences will be daunting. In addition, fracking and its benefits tempt us to pursue the use of fossil fuels and delay the development of renewable energy sources. This commits us to worldwide warming and to a continuing need for climate-change adaptation.

4. We're going to increasingly find ourselves in a zero-margin world, because we will hit the limits of sustainability. This will take two forms. First, in the developed world free markets and competition will lead us to this circumstance. Take inventory. Historically, manufacturers and retailers would amass substantial inventories of their products to meet unanticipated demand. No more. Supermarkets used to have weeks of canned goods and other supplies on hand. Today, delivery trucks bring to the back door what's needed to restock shelves just as customers leave the main entrance with their purchases. Electric utilities used to be able to meet customers' peak demand locally. This meant they all had excess generating capacity to draw upon. No longer. Today, thanks to deregulation, they meet peak demands by purchasing power off regional and national grids. Airlines used to fly planes with dozens of empty seats. Today they fill those seats through marked-down, last-minute sales. Margin in any form is considered waste.

It is vigorously stamped out wherever it appears. It has been replaced with zero-inventory, just-in-time approaches to manufacturing and services. As a result, when things go wrong, when a snowstorm stops airline flights and deliveries of food to stores, when ice storms down power lines or a heat wave drives electricity demand above the national supply, the disruption is immediate and can be severe.

In the developing world, declining margins are imposed by the developed world and the markets. Again, there was a time in all countries, no matter how small or undeveloped, when food that was needed for local consumption was grown locally. But today, in such countries, food grown for export (such as coffee and bananas) can achieve a greater monetary return than food grown for domestic consumption. Foreign countries and globalized corporations negotiate with local farmers for production of goods to be sold worldwide. This is a dangerous situation. We'll experience shortages of all sorts of key resources, and this experience will be highly localized and episodic and highly unevenly distributed across peoples and nations. Recent experience in grain markets has shown this. As the United States and other countries have switched grain production from use for direct human consumption or for feeding livestock to feedstock for biofuels, the resulting price shocks have lifted staples such as rice and maize out of range for the world's poor. We'll likely see other manifestations of this trend throughout the century.

5. Finally, we will find that top-down, command-and-control approaches to dealing with all of these challenges will tend to be ineffective at best and dysfunctional at worst. The rate of change is so rapid that we will constantly try to solve tomorrow's problems with yesterday's capabilities, because of the failure to internalize the real meaning of social and technological change and its import for us. And the people who are in charge of the that top-down, command-and-control model are generally the oldest and thus the most prone to be operating on worldviews that used to be cutting edge but that are no longer relevant.

This is the business-as-usual scenario or persistence forecast for the remainder of the 21st century. This is the kind of real world that is likely if we take no action.

4.4. Alternative Futures

It doesn't necessarily follow that our persistence forecast should prompt action. It might be that this likely world is precisely the future we all seek or the best we can hope to achieve. Before we contemplate steps we might take in response to this forecast, we might therefore ask what kind of world do we *want*?

Or, more precisely, what kind of *real* world do we want?

This last qualifier is important.

To see this, let's return to each of our five trends. Start with climate. Over the course of Earth's history, weather patterns have been resolutely variable, over all spatial scales and over every time horizon, however short or long. It happens that weather patterns have been relatively stable for a couple of centuries and that this corresponds to the time over which climate has been of interest as a concept.

Consider two possibilities. The first is that this stability of weather patterns with the success of the human race is simply an accident of timing— that it just *happened* climate was stable at a time when we'd developed to the point where we were able to grasp the notion of climate itself. The second is that if weather patterns hadn't been stable for several hundred years, the human race would never have been able to progress to the point we have. Nature would have been continually frustrating our early efforts at agriculture, for example, and we'd never have left our hunter-gathering or our nomadism.

In either instance, our future choices and options do not necessarily include a stable climate. Many times, the popular press insinuates that if we only manage to reduce or stabilize our carbon emissions the climate will in turn remain as it is today.

That is not the case. What we do with respect to carbon emissions, especially from fossil fuels, will indeed contribute to our climate future, but such human decisions and actions are not the only determinants of climate. The same natural variability that has cycled the planet between ice ages and warmer periods will continue. Our choice going forward is not between "climate change" and "no climate change." Rather it is a question of what contributions we make to the variability on all time scales. It's not a case of "getting the genie back in the bottle." The genie was never in the bottle in the first place.

What corresponding statements can we make about extremes? Realistically, we can't eliminate them. Earth does much of its business through extremes. In fact, what we call climate is in effect the average of extremes of heat and cold and flood and drought from place to place and over long time scales.

In a way, this both simplifies and complicates any effort to attribute such extremes to climate variability, as is often done. These days, whenever we have a hot, dry summer, as we did in the United States in 2012, or a particularly devastating storm, such as Hurricane Sandy of that same year, we see a burst of news stories wondering whether (or the extent to which) the events can be attributed to climate change. The answer is both "of course" and "no." Of course, because those events, when summed, will define the averages. No, because such events have occurred under all climate regimes.

But, even if we could eliminate such extremes, we probably wouldn't want to. They play their essential role in making the planet habitable. For example, they reduce the temperature differences between equator and poles, and they provide much of growing season rainfall in our agricultural belt. We and the planet have grown up together, as it were. We've become accustomed to each other's foibles. The Rocky Mountain snowfalls that threatened our ancestors during the Gold Rush years immediately after 1849 today enable the winter skiing industry that is so lucrative for the states of Colorado and Utah. The El Niño rain- and snowfalls that plague California at the same time top off the northern California reservoirs that supply much of the water needed for the state's agriculture and for the population of Los Angeles to the south.

As for the unintended consequences of human success, again our goal is less about eliminating them and more about anticipating them, seeing side effects and consequences of our decisions and actions more completely and well in advance, giving us time to reduce their negative implications. Some claim this has been accomplished with fracking, for example, while others say negative implications have not been reduced or eliminated at all.

As for a zero-margin world, a similar statement applies. If we can create entities and ways of doing business where margin and capacity ebb and flow in response to need, then we will have achieved enough. Here's a simple example of what can be done, given notice. The U.S. Department of Commerce is charged with carrying out a national decennial census. The Census Bureau's staff totals some 4,000 professionals. But during each decennial census, the Bureau hires something like 500,000 temporary, part-time employees. Every Christmas season, retailers add 600,000 people to the temporary workforce. To the extent this kind of flexibility and adaptability is incorporated into the ways companies and government institutions do business, low margins aren't a problem. But in other areas—flu vaccinations, for example, or the supply of heavy equipment such as the big substation electrical transformers vital to the regional power network, which can be damaged by space weather—the picture is not so sanguine. As diseases such as West Nile virus emerge, or as flu strains mutate, we're periodically reminded that the supply of vaccinations is falling behind. And there's a 90-day backlog for the biggest electrical transformers. Or look at world grain stocks, which are variously measured in tons or in terms of days of consumption. The latter measure fluctuates but averages out at about 100 days.

It's probably more realistic, therefore, to frame our goals less in terms of building margins per se and more in terms of building resilience and growing more adaptive.

Which brings us to top-down, command-and-control decision-making. Certainly societal goals should include building the effectiveness of such decisions, but at the same time (and this is the thesis of much of the book) the goal should be to distribute decision-making broadly, putting and keeping it close to where actions are required.

4.5. Recap

To review, the kind of real world we want and can achieve is one where:

- we anticipate climate variability and continually adapt our decisions to be optimal with respect to climate that is coming;
- we anticipate and build resilience to extremes of nature and reduce any threat they may pose;
- we anticipate and accommodate the full range of side effects of our major decisions and actions, rather than allow these to catch us by surprise;
- we accept that margins will be small but develop an ability to adaptively expand and contract in response to changing needs; and
- we distribute decision-making widely, placing it close to action sites.

How do we get there? This is the key challenge for 21st-century humanity.

How do we move from our current trajectory to one more likely to ensure our survival (or just more to our liking)? All of these challenges need not overcome us. We can meet them, transcend them, and make a better future for ourselves as past generations have done.

But we face constraints. To begin, meeting these challenges is not simply a matter of implementing what we know. All we know is a little bit about what *doesn't* work, based on limited past experience. And even that knowledge is fuzzy and incomplete because we haven't approached our task of living on the real world as a scientific problem. True enough that we've collected a little data, formed some hypotheses, and tested some ideas, but only in the most haphazard, intermittent way. We haven't been at all disciplined.

What's worse, we feeling a bit financially straitened right now. We don't have unlimited resources to throw at these problems.

And finally, and perhaps most distressing, we don't have all the time in the world. The atmosphere, the oceans, and the land surface and biota are all responding to our actions on their timetable, not ours. Extreme events—volcanic eruptions and earthquakes and floods and drought—are queuing up on a schedule not of our choosing or knowing.

We need approaches that are *cheap*, *quick*, and *effective* (or can at least get more effective fast). It might seem that approaches meeting those three criteria simultaneously might violate some law, that of entropy, for example.

But the reality is there are four tools available in our 21st-century world that would seem to meet these standards if not individually, at least in aggregate. Taken together, they just might work:

1. *A basis of facts.* Without Earth observations, science, and services, we're essentially flying blind into the future. But as we build our ability to monitor atmospheric, oceanic, and land surface conditions, and as we gain an understanding of how the earth's natural systems function and their interdependences, we can predict how conditions will change. More importantly, we'll be able to tease out and distinguish between how environmental conditions are poised to change on their own, and how they are changing in response to human decisions and actions.

2. *Policies.* The state and trends of the natural earth, though intrinsically interesting, matter in much more practical ways. They impact us as individuals and as a society, posing economic opportunities, revealing environmental responsibilities, and threatening our safety and health. And for this reason, we're constantly anticipating what the real world we live on will do next, and we're scrambling to respond when it goes off-script. The changes are so important, so mish-mashed with one another, and occurring so rapidly we can't afford to respond to each new situation de novo, from basic first principles. Instead, we've got to act on the basis of public policies. These leverage our intelligence. They equip us to make better decisions, more quickly and more uniformly, in much the same way as high-level languages like FORTRAN enable us to program more swiftly and accomplish more than we could working in so-called machine language, directly with the 1s and 0s that form the essence of the digital world.

3. *IT and social networking.* If our public policies presently in force for resources, environmental protection, and hazard resilience were spot-on, we'd need no more tools. But in fact, all our policies could stand some improving, and many of our policies are seriously deficient. To make a success of real-world living we need to hone public policies in the first category and trash those in the second, replacing them with more effective alternatives. Here

we face two challenges: (1) out of the vast array of possible policy choices, we know only dimly a priori which ones might be helpful, and (2) we don't have all the time in the world. Fortunately IT and social networking offer the promise of rapid exploration of policy alternatives, learning from experience and innovation.

4. *Local-leaders.* Mere head-knowledge of the right things to do is not enough. We must match that with action. If top-down, command-and-control leadership is no longer effective, we need to replace and/or augment it with something else. One promising alternative enabled (or perhaps even demanded) by IT and social networking is pushing the theater of action out to the local level, while maintaining a connection to high-level leaders.

We explore these four tools in the 21st-century toolkit in the next few chapters.

5

A BASIS OF FACTS

False facts are highly injurious to the progress of science, for they often endure long; but false views, if supported by some evidence, do little harm, for everyone takes a salutary pleasure in proving their falseness. —*Charles Darwin,* The Origins of Man, *Chapter 6*

It ain't what you know that hurts you, it's what you know that ain't so. —*widely attributed to Mark Twain, Will Rogers, Satchel Paige, Yogi Berra, and others; thus, author unknown.*

Here are three falsehoods:

1. Climate is unchanging.
2. The assimilative capacity of the atmosphere is essentially infinite.
3. Weather is unpredictable.

Looking through the lens of history and the last century of science and technological advance, these three statements appear to be mere views, but for the hundreds of thousands of years of human experience prior to that time they were treated more as facts or truths. Certainly, they've been the prevailing mindset for some time and persist even today, much as the idea that the world was flat endured long after most scholars had come to a different conclusion.

Over the past 100 years or so, meteorologists haven't merely tweaked or refined these long-held notions. They've turned each on its head.[1] Today, for example, we know that throughout the geologic time spans, Earth's climate has shown itself to be resolutely variable. In fact, had it not been for the comparatively steady (and perhaps relatively unusual) conditions of the past several hundred years, we might not even have the notion of a climatological average and seasonal to inter-annual departures from it. In other geologic periods, such an idea might have been of less practical use.

You and I also know, because we live in today's Age of Nonlinearity, that the atmosphere's assimilative capacity is finite. It's constrained even on global scales. We can't empty pollutants into it willy-nilly without consequences. Worldwide, some three million people a year die prematurely because of local or regional air pollution—one million deaths a year in China alone. A century or so ago, as the world's peoples stepped up consumption of fossil fuels and greatly expanded smelting and manufacturing at the start of the Industrial Revolution, they were soon dismayed to experience considerable local environmental damage in the immediate vicinity of smokestacks. They quickly determined that by increasing the stack height and releasing the pollutants at heights well above the surface boundary layer they could "solve" the problem. They maintained this new engineering practice for decades with seemingly good results. Eventually, however, everyone began to notice signs of environmental degradation once again, but now across entire regions, not just locally. In the United States, for example, this showed up as acidification of lakes and streams across the Northeast. The trees in the forests and the fish in the streams began to die.

Environmentalists were concerned. By contrast, industrialists were reluctant to acknowledge the problems and shoulder the corrective costs and measures they would imply. The effects and the arguments were transboundary. Canadians complained their environment was suffering because of U.S. industry and the prevailing winds. In Europe, the Scandinavians pointed an accusing finger at German industry. The public outcry and industry resistance gave birth in the United States to the 1980–1990 National

---

1. As an aside, this suggests that over this period, meteorologists have more than earned their keep, just for these achievements alone. It is something to keep in mind when it comes to the Government Performance and Results Act (GPRA) as it applies to Earth observations, science, and services.

Acid Precipitation Assessment Program (NAPAP). Proponents said this program would nail down the science. It would inventory emissions of acid precursors, look at transport and diffusion and impacts, and sort out pollutant source-receptor relationships. Environmentalists in this country and in Canada saw NAPAP differently. To them it seemed a mere delaying tactic. The 10 years of work informed the 1990 reauthorization of the U.S. Clean Air Act, which eventually led to adoption of scrubbers and other technologies to reduce emissions of particulate and acidic precursors at smokestacks, as well as cap-and-trade policies on emissions, in order to foster such cleanup.

Of course, throughout the 1980s and since, the use of fossil fuels has continued unabated, and has even accelerated. Today the use of fossil fuels is so widespread as to be causing global-scale changes: temperature increases, the melting of ice caps and glaciers and concomitant sea-level rise, acidification of entire oceans, and changes and migrations in land and marine ecosystems. And the acid rain debate, which throughout the 1980s had seemed so rancorous, seems mild in hindsight. It seems to have been the climate-change battle on training wheels, while the climate-change battle has proven to be the acid rain debate on steroids.

Which brings us to the third falsehood. Even as the great 19th-century American writer and humorist Mark Twain was supposedly opining that "everyone talks about the weather, but nobody does anything about it," his contemporary Cleveland Abbe[2] and other scientists of the time had begun issuing daily weather forecasts at the U.S. Army's Signal Service and formulating weather prediction as a problem in physics. Since then, with the addition of each new observing capability (surface networks and weather balloons, weather radars and satellite-based instruments), as well as the invention of digital computers and the development of numerical weather prediction, weather forecasts have steadily improved and grown more useful. It used to be that a missed weather forecast would bring smiles, a clucking of tongues, and a joke or two. Today's expectations and stakes are higher. To start, it takes less error to constitute a "miss." Abbe's 1880s-vintage Signal Service divided the eastern half of the United States into nine regions; for each region, a forecast of "rain" or "no rain" was assigned. A "rain" forecast was considered

---

2. Cleveland Abbe (1838–1916) was an important and underappreciated meteorologist of the late 19th and early 20th centuries. My historian friend and colleague Edmund Willis and I published a brief biography of the man. You can find it here: Willis, Edmund P., and William H. Hooke, 2006: Cleveland Abbe and American meteorology, 1871–1901. *Bull. Amer. Meteor. Soc.*, **87**, 315–326.

a success if it rained anywhere in the region in question at any time over a 24-hour period. A little over a century later, in the winter seasons here in the Washington, D.C., area, for example, the complaints flood in whenever forecasts have missed the boundary separating snow from rain by as little as 30 miles or so (essentially the distance separating downtown D.C. from its far-western and northern suburbs), even when the forecasts have characterized the associated uncertainties rather well (again as compared with standards prevailing only a few decades ago).

Humanity has paid a substantial price for failing to expose these falsehoods earlier.

5.1. The Costs of Not Knowing[3]

As the saying goes: "It seemed like a good idea at the time."

Getting the facts wrong, being ignorant of the facts, and flying blind into the future entails huge risks and, on occasion, incurs staggering economic and other societal costs. One way to see this is to look back on some of our real-world choices, decisions, and actions. Hindsight is 20/20, even though each choice seemed like a good idea at the time.

Perhaps we can start with dams. Historically, damming rivers has served many useful purposes: flood protection, hydroelectric power, water-resource storage for agriculture, recreational opportunities, and more. But based on U.S. experience, we know today that building dams on watersheds to realize such societal goals carries unintended consequences: habitat loss; ecosystem changes; sediment buildup behind the dams requiring constant dredging; and sediment loss downstream from the dams leading to loss of coastal wetlands and deltas, often far downstream. As a result, our national enthusiasm for such infrastructure, dating back a century or more, has dimmed. But suppose we'd known back then what we know today. Suppose we'd been more astute in identifying these unintended consequences and their costs. We might well have not built many of those dams. We might have built them in different locations or in different ways. We might have saved ourselves a lot of difficulties we now face trying to maintain ecosystems that had once thrived on the unmanaged, free-flowing rivers. We could have been quicker to identify more effective and more enduring strategies for coping with riverine hazards and benefits. We may well have retained many of the ecosystem services, alternate tourism and recreational opportunities, and other

---

3. Excerpted in part from a post on LivingontheRealWorld, dated March 13, 2013: "What did we know . . . and when did we know it?"

economic benefits afforded by floodplains. We might not be experiencing large repetitive flood losses.

Domestically, we think we've learned this lesson here in the United States. New watershed control projects entering the works usually maintain what's in place or take a softer form. In some locations, as along the Connecticut River and elsewhere in New England, dams actually have been or are being removed (more about such retrofits later). But internationally, China is today working on nearly 300 major dam projects, including many across Southeast Asia intended to provide electrical power not just to those countries impacted but also to China. Host countries such as Vietnam, Cambodia, and Myanmar (Burma) are beginning to question these projects. But the development is still going ahead. Brazil is pursuing numerous major dam projects with vigor across the Amazon. And in these tropical locations, dam reservoirs can release large amounts of methane; Brazil's Balbina dam, built in the 1970s, emits more greenhouse gases than an inefficient coal plant of comparable generating power.[4]

Similar statements apply to our coastal zone management nationwide. Think of how different the impacts of Hurricanes Katrina and Sandy would have been if in the previous century we had prized undeveloped coastlines and located our coastal businesses and residences a little (not really that much!) farther inland. Imagine a largely pristine American coastline, where with the exception of ports and harbors needed for international trade the shoreline per se retained its full natural splendor. And imagine just inshore, within the same kind of easy commuting distance from nearby hotels to the beach that people enjoy when staying at ski lodges or Disney World, we would find large numbers of tourists and the hospitality industry supporting them. We could have preserved the priceless marine resource and recreational value of the coastal zone while simultaneously maintaining or even enhancing its economic contributions and reducing costly losses to coastal hazards.

Our worldwide dependence on fossil fuels poses the largest challenge of all. Were it easy to unwind the trillions of dollars that have been invested in the infrastructure needed to extract energy for business, homes, and transportation from fossil fuels, we might well be already far down that road. Suppose, years ago, as we had embarked on the development of 20th-century energy resources, we had known what we know today about the cumulative effects of burning coal, oil, and natural gas in today's quantities. Suppose

---

4. According to *The Economist*, May 4, 2013.

we had done the energy R&D needed so that alternative technologies had been at hand. We might well have created a richer set of options and chosen a different direction.

But we didn't know what we most needed to know at the time.

Could such investment errors have been averted? Would it have been possible to accelerate the pace of science to the point where the needed knowledge and understanding could have been amassed in time to go in a different direction? Would that alternative direction have proved much better? Alternatives necessarily bring unintended negative consequences of their own. Would these have been better or worse?

Would it have been possible to accelerate the pace of science to the point where the needed knowledge and understanding could have been amassed in time to go in a different direction? Would that alternative direction have proved much better?

All these are hypotheticals; we can never know for sure. Living on the real world is not a controlled experiment.

But we do have some real-world illustrations of the positive effects of learning from the experience of others, and leapfrogging technologies to national benefit. Perhaps the best publicized example comes from telecommunications. U.S. and European infrastructure rely heavily on landlines, put in soon after the invention of the telegraph and telephone, a century or more ago. By contrast, in the developing world, most notably Africa, economic growth is being jump-started by the extensive use of cellular phone technology, which relies on a much cheaper, more quickly installed infrastructure of transmitter towers. And the current economic boom in China and India stems in part from their ability to build a newer, more modern infrastructure to support their economies.

It's analogous to playing chess against the real world, but instead of thinking several moves ahead, we're realizing each move's consequences only after lifting our hand from the chess piece at each turn. The conclusion is that we can never know all the ins and outs of Earth and the related natural sciences soon enough or well enough. When we make a move, we're not adequately accounting for what Earth will do in response. And like that chess neophyte blundering through the game, we set ourselves up for repeated losses.

How much this concerns us varies. A colleague of mine has ventured that as a group, economists (apart from a small handful valuating ecosystem services) are relatively unconcerned with climate change because they see most of the impacts as ameliorated in part by possibilities for substitution (e.g., a hotter world just calls for a bit more air conditioning). By contrast, he suggests, ecologists see climate change as a catastrophe because they're the ones daily discovering new sensitive interdependences of all flora and fauna and witnessing and documenting the myriad contributions of biodiversity to the ecosystem function and services we enjoy today (some economists of course explore this space as well). Ecologists more than anyone confront on a daily basis how disruptive climate change can be to these natural processes. (My colleague gauges physical scientists to be somewhere in the middle.)

We cluck our tongues as meteorologists struggle with uncertainties in their forecasts, but by comparison the uncertainties in ecology are immense. Are our ecosystems so fragile that the slightest disruption compromises them forever? Or is the opposite true? Do they typically embody so many redundancies and compensating mechanisms as to be insensitive to climate variation and other changes? (Whether robust or not, ecosystems are the product of millennia of natural variability.) It's likely that the answer varies from ecosystem to ecosystem, and question to question. Some of all of this is going on, in different ecosystems, in different places, with different consequences. But the simple truth is that we don't know.

Surely it behooves us to be less vague, which requires more investment in Earth observations and science. Will we keep learning what we need to know only in hindsight? Or will we get better at seeing the consequences of our decisions and actions in advance—before bad things happen—on a scale that makes the costs of our previous mistakes seem trivial by comparison? Will we become true chess players?

We've looked briefly at three examples here. But there's much more to each story, and there are many more stories: invasive species; land use, in contexts ranging from urbanization to agriculture in semi-arid areas; and management of groundwater. You can no doubt add your own.

Right now, with respect to Earth observations, science, and services (Earth OSS) and what they're telling us about the future for (1) water, food, and energy; (2) protection of the environment and ecosystems; and (3) resilience to hazards, mankind is essentially flying blind into a problematic 21st century. To continue the metaphor: To present, we've been flying well over the plains, under visual flight rules (VFRs) conditions. We've had a relatively simple, twofold problem, and we could see everything we needed to know. But now, when it comes to what Earth will do next and its carrying capacity, relative to the seven billion of us, we're in a cloud, the wings of the aircraft are icing, and we're below the altitude of the mountains, hoping to find a break in the clouds and a safe place to land. Conditions are instrument flight requirements (IFR). We need instruments that will show our location, the terrain, the aircraft altitude, and much more.

The issue really is:

What *will* we know, and when *will* we know it?

5.2. The Value of Knowing

If the *costs* of not knowing are large, what can we say about the *value* of knowing?

Political leaders ask scientists such questions all the time. It's only natural. In fact, what they really want to know is the return on investment (ROI) for any and all federal expenditures and associated legislation and regulation: that is, the amount of economic benefit from such expenditures and the timing of that benefit. The larger the benefit relative to the cost of any legislative measure (appropriations, regulation, tax, etc.), the better; the sooner that benefit is realized, the better. Benefits that come in only after a delay of several years are heavily discounted, and the more so the higher the prevailing rate of inflation. Know these returns, and perhaps the policy process (see Chapters 6–8) is correspondingly simplified.

Of special interest here is the ROI on Earth OSS, although useful information on such value has proved hard to come by. Often the observing technologies and the research being funded are unproven. How well they work will combine with social change to shape how Earth OSS are used in the future. That's probably the biggest contributor to the uncertainty. Some benefits are more readily monetized than others. Research on valuation is underway but in its infancy.

Some economists might choose to remind us that in free markets, we don't necessarily need to calculate the value. We can ascertain value from simply inventorying how much is spent, and infer that the annual benefit is comparable to that annual expenditure and/or that the marginal benefit from that last increment of investment is roughly equal to its marginal cost. We don't currently enjoy a reliable inventory of these investments, and many investments in Earth OSS serve multiple purposes. But we can gain a rough idea, an order of magnitude, by building a number in the following way. Within the United States, we spend $5 billion per year on the National Oceanic and Atmospheric Administration (NOAA); another $2 billion per year on the Earth sciences in the National Aeronautics and Space Administration (NASA); and $1 billion per year each at the National Science Foundation

(NSF), the Department of Energy, and the U.S. Geological Survey. Add a few billion dollars more for research at the U.S. Department of Agriculture and the ecological science and services at the rest of the Department of Interior, and we arrive at a total something like $15 billion per year. (This estimate is highly uncertain, but it tells us that the U.S. investments are unlikely to be $150 billion per year at the high end or $1.5 billion per year at the low end.) To extrapolate to a global figure, let's take the ratio of U.S. gross domestic product (GDP) to world GDP. We'd then get a global annual investment in Earth OSS of something like $70 billion per year.

These figures amount to about 0.1 percent of U.S. and global GDP, respectively.

Some might therefore be tempted to argue that all the Earth OSS contribute no more than a $70-billion-per-year to a $70-trillion-per-year global economy. But it might be that the $70 billion figure was not arrived at by any cost benefit analysis but rather merely represents the accidental result of past history. Furthermore, it might be that as a consequence of the uncertainties in the estimates of value, Earth observations and their use have been determined as much by what is easy and/or inexpensive to measure, and levels of investment in prior years, as by any more rigorous or logical criteria. If ground-based radar wind profilers had been developed before weather balloons, for example, instead of decades afterward, we might be using the latter today. As it is, radar wind profilers remain largely a research tool, available only at certain locations and for limited periods. Similar examples abound: laser wind sensors, chemical samplers testing particulates and gaseous pollutants, satellite constellations for measuring humidity worldwide by occultation, phased-array radars to replace their mechanically steered predecessors, and so on.

There is some evidence that the federal funding for Earth observations over the years has not been determined by a well-founded analysis of the ROI but instead, on historical levels, incrementally adjusted. For many of my later years in NOAA during the 1990s, the process for funding our agency's observations went something like this: The NOAA budget was part of a larger Commerce, Justice, and State Departments appropriation that in turn was capped along with all other government budgets by the Gramm-Rudman-Hollings Balanced Budget Act. What resources were assigned had to be allocated across the departments. It was a zero-sum game. During much of the period, the State Department needed sharply rising budgets to upgrade the physical security of its embassies and consulates and their staff around an increasingly troubled and dangerous world (these concerns were dominant, even in a pre–9/11 context). The Justice Department was progressively more deeply embroiled with drug trafficking and illegal immigration across U.S. borders. Lacking any rigorous analysis on the dollar benefits of knowing future weather or the dollar losses associated with not knowing, Commerce struggled to retain budget parity in this environment.

Today a similar zero-sum game continues to threaten NOAA. Satellite platforms and instruments provide one of the few viable options for getting Earth observations of the needed detail and global coverage. But the costs of those platforms are rising, and the nature of big, high-tech government procurements perversely increases risk of programmatic delays and associated cost overruns in satellite construction and launch. In early 2013, the Government Accountability Office added both the NOAA polar-orbiting and geostationary-orbiting satellite programs to its biennially published report titled "High Risk List": the list of consequential government programs at some jeopardy of failing in their intended purpose. At the same time, zero-sum budgeting means that other NOAA programs, such as fisheries research and management and coastal zone management, are compromised in turn by the satellite woes.

The lack of a solid analysis of the ROI for these programs contributes to the instability and uncertainty. In its absence, there's no ability to say that "the costs (of the program) have risen, but they're still dwarfed by the stream of benefits we can expect at the other end . . ."

The *costs* are difficult enough to measure; they're spread over multiple years and numerous offices, and they reflect work across both the government per se (NASA as well as NOAA is involved, as well as the Department of Defense [DoD]) and a number of private-sector contractors and subcontractors.

But the *benefits* are characterized by far greater uncertainties. What are the benefits of weather science and services to the agricultural sector, say? Or to energy utilities? Or water-resource or transportation managers? And what is the contribution in turn to those services from, say, satellite platforms and instruments? Are such platforms essential to such services, or do they merely provide marginal additions to benefits? What about nonmonetizable benefits, such as improvements to public safety in the face of weather hazards, or the contributions to geopolitical stability worldwide? Are the benefits true public goods? Or are they conferred inequitably among a select few?

We're nowhere close to answering these and other such questions or even formulating the list of issues that require such answers. Much of the analysis extant is anecdotal and confined geographically and temporally, as in: the benefits of weather data and forecasts to peanut growers in a single Virginia county; the expected wind energy available from turbines sited in the San Gorgonio Pass of California; or the reductions in the number of malaria cases among Nigerians.

Most seriously of all, however, investment levels in Earth OSS have been set looking in history's rear-view mirror instead of looking ahead to what we'll need from Earth OSS in coming years, as we struggle to cope with climate variability and changes, hazards, and environmental degradation in a zero-margin world.

Therefore, chances are good that because we lack credible policy analysis and economic valuation, we are greatly underinvesting in Earth OSS. In the absence of hard data and in the belief that hard data will be of less use to decision-makers than a better conceptual framework, consider four levels of knowing.

1. *Documenting human failure after the fact.* Scientists particularly should be concerned about this one. Sometimes, in more cynical moments of thought, it seems that our community has been marketing ourselves along these lines for years. We say something that unintentionally sounds like this: *We're in the business of documenting human failure. That is, we're doing a lot of research on (1) how production and consumption of water, food, and energy are on an unsustainable path; (2) how human activity is destroying habitat, denuding the landscape, reducing biodiversity, and degrading the environment; and (3) how losses to hazards are inexorably rising faster than inflation. But lately, the pace and scope of human failure is accelerating. It's at the same time growing more complex. If our efforts to document are going to keep pace, we'll need more funding.*

This is not the best marketing approach to a world more in need of answers than in need of criticism. The value of after-the-fact knowledge, however richly detailed, is limited.

Unless there is the second level of knowing.

2. *Learning from experience.* This is what happens in the world of transportation. The National Transportation Safety Board works with other stakeholders (the U.S. Department of Transportation, as well as its state government counterparts, vehicle and plane manufacturers and operators, and so forth) to ensure that accidents translate into lessons learned. By contrast, this is not what happens when it comes to natural hazards. In 2001, a famous trio of social scientists—Gilbert White, Ian Burton, and Bob Kates—offered a different perspective, and explanation, for what is going on there. In an article titled "Knowing better and losing even more: the use of knowledge

in hazards management,"[5] they concluded that one or more of the following could be occurring: (1) knowledge is lacking, (2) knowledge is available but unused, (3) knowledge is available but used ineffectively, (4) there is a time lag between the application of knowledge and the results, or (5) the best efforts to apply knowledge are overwhelmed by the rapid increase in vulnerability. (Their article goes into a lot more detail and is worth a serious read. This topic is explored further in Chapter 7.)

Real learning from experience ought to be reflected in improved, more effective decisions with respect to impending hazards to public safety, resource shortages, and threats to the environment. These might take two forms. First, Earth OSS ought to be useful for the third level of knowing.

3. *Guiding warnings and emergency response.* Often, surveillance of water, energy, and agricultural resources or weather observations of developing and approaching storms are used to warn the public dependent on such resources or in a storm's path.[6] The Hurricane Sandy forecast described in Chapter 1 illustrates the point. Hurricane Sandy caused loss of life. The losses were greater than they might have been; the warnings based on the weather conditions might have been reframed in ways that would have prompted a more effective public response. But the fatalities were also far lower than they could have been had the forecasters failed to predict the westward turn of the storm toward New York and New Jersey, catching everyone unawares. However, many of those who evacuated and saved their lives nevertheless lost their homes, valuable property, treasured keepsakes, and in some cases their livelihoods because the knowledge was not used more completely.

Used more effectively, knowledge on the climatology of such storms and the risks they posed over time might have led to the fourth level of knowledge.

4. *Avoiding natural catastrophes well in advance.* At this level, knowledge has far more value because it's used to guide land use and building codes, as well as the development and deployment of critical infrastructure such as roadways, electrical power grids, water and sewage systems, communica-

---

5. White, G. F., R. W. Kates and I. Burton, 2001: Knowing better and losing even more: The use of knowledge in hazards management. *Global Environ. Change,* **3B,** 81–92.

6. Kevin M. Simmons and Daniel Sutter have provided an interesting analysis of costs and benefits for tornado warnings in the larger context of the overall impacts of such hazards: *Economic and Societal Impacts of Tornadoes,* American Meteorological Society, 2011.

tions, and even hospitals. Something like this happens in the Rocky Mountains, which once posed a deadly hazard to settlers crossing the country from the Midwest to California but where today ski resorts are a major wintertime recreational destination.

It should be clear from these four scenarios—(1) documenting human failure, (2) learning from experience, (3) guiding warnings and emergency response, and (4) avoiding natural catastrophes well in advance—that the benefits of Earth OSS are not fundamental physical constants that can be measured once and for all and then forgotten. They fluctuate wildly depending on national policies. Here's a sample comparison:

There was a time when electricity was generated locally for local consumption, everywhere across the United States. As a result, individual electrical utilities had to carry the capacity needed to meet peak demand, usually experienced during the early evenings of very cold, short winter days or the very hottest of summer days when air-conditioning usage was high. This meant that most utilities had idle surplus capacity most of the time, representing significant cost. In a large country such as the United States, different locations are rarely experiencing peak demand simultaneously. So, over time, the nation developed huge regional grids for moving electrical power around the system and at the same time deregulated the industry. Utilities are now free to reduce their excess capacity, and as local, intermittent needs arise they simply purchase extra power from the grid on the so-called spot market. In response to the reduced margin in the system, the value of weather forecast support to the industry and the power-consuming public has increased substantially.

By contrast, dam operators on U.S. watersheds have historically controlled storage and outflows from their respective dams in accordance with extensive regulations based solely on water levels in the reservoirs behind those dams—not on any outlook for dry weather or heavy rain over the watershed for the next few days or weeks. As a result, the value of such environmental forecasts to water-resource managers is forced to be close to zero.

To realize this highest level of value from Earth OSS therefore requires not just having the knowledge, but using it, and using it effectively. And how we use such information is determined largely by policy. We'll tackle this in the next three chapters.

But before turning to that subject, we have more to consider with respect to the facts themselves. For example, there's the question of what we need to know.

5.3. What Do We *Need* to Know and Measure?

It turns out that we need to know and measure just about every aspect of real-world conditions. It's equally important, and indeed urgent, that we know *how those real-world conditions will change*. Knowledge and understanding are the keys to anticipating at any given point what Earth itself and its atmosphere and oceans will do *next*, not only because of internal processes and dynamics but also in response to human intervention.

Recall, as discussed in Chapter 2, that Earth's atmosphere and oceans are *chaotic* systems.

As a result, meteorologists always crave more data. They want to measure everything, everywhere, all of the time, with pinpoint precision. That obsession has been captured in doggerel:[7]

More data, more data,
Right now and not later.
Our storms are distressing,
Our problems are pressing.
We can brook no delay
For theorists to play.
Let us repair
To the principle sublime:
Measure everything, everywhere,
All the time . . .

. . . More data, more data,
From pole to equator;
We'll gain our salvation
Through mass mensuration.
Thence flows our might,
Our sweetness, our light.
Our spirits full fair, our souls sublime:
Measure everything, everywhere,
All the time.

---

7. Poem by A. Fleisher. Originally published in 1957 in the *Proc. Sixth Weather Radar Conf.*, American Meteorology Society, Boston, MA, p. 59. Slightly modified by Peter Black, NOAA's Hurricane Research Division.

It shall come to pass, even in our days,

That ignorance shall vanish and doubt disappear.

Then shall men survey with tranquil gaze

The ordered elements shorn of all fear.

Thus to omniscience shall we climb,

Measuring everything, everywhere, all the time.

(There's more to this poem. Unfortunately, there's *considerably* more, and in its entirety it's excruciatingly painful, and by now you surely get the idea.)

Back to what we need to measure. Suppose, for openers, we limit ourselves to the atmosphere and the task of global numerical weather prediction (NWP). To initialize our model, we would ideally specify the pressure, temperature, humidity, and the east–west, north–south, and vertical components of the wind—everywhere worldwide, at all locations, and at all heights above the ground. This information is needed to solve the several equations that govern atmospheric motions, which we must solve *simultaneously*. (These equations are too complicated for us to solve exactly, by the way. We can only approximate the solutions for brief intervals using arithmetic—more on that later.)

It's difficult enough to make the measurements at any one spot, particularly above Earth's surface. But the desired global coverage—getting the measurements everywhere with the fine spatial and temporal detail required—is truly daunting. For much of the early history of meteorology, data were collected largely from instruments in sparse networks, randomly scattered across the world's land,[8] but somewhat more concentrated in or near population centers. Weather balloons improved the situation somewhat, but these have never been available for more than about 100 sites worldwide, mostly over land as well. In recent years satellite-based soundings have provided dense coverage of some parameters over both land and sea, and from pole to pole. Such advances and improvements have been marked but were hard won.

That's just for the task of weather prediction. Climate modeling is easier in some respects (it's straightforward to predict that *the coming winter will be colder than the present summer*, for example). But there's additional in-

---

8. Indeed, the early numerical weather forecasts weren't global but hemispheric, confined to the Northern Hemisphere, which is largely land, and not even attempting to treat the Southern Hemisphere, which is largely ocean.

formation that's essential: the location of clouds and the cloud microphysics (what's going on in the cloud to make droplets or ice crystals); details of the atmospheric gaseous and particulate composition so that we can describe the radiative transfer of heat from any layer of the atmosphere to the air below or above; underlying conditions on sea and land, including changes in land-use patterns; and much, much more. Thousands of books and millions of journal pages have been dedicated to describing all of this.

But remember, the 21st-century task we face also demands that we keep track of the world's resources. Where will we get all the energy we need? Where will the fresh water and the food and the fiber come from? What will be the accompanying effect on landscapes, land use and habitat, biomass and biodiversity, air, water, and soil quality? Much of the resource data demand measurements at or below Earth's surface or in and beneath the world's oceans. Satellite-based platforms are of limited use. And ecological measurements are particularly labor-intensive. Then there's the nontrivial matter of monitoring the sun and understanding its associated dynamics, including the resultant space weather, and tracking asteroid threats.

Whew!

Underlying all this is the need to understand at a conceptual level how all the bits and pieces and the processes and trends of the puzzle that is the Earth system fit together. What's the *nature* of Nature? How does Earth, its atmosphere and oceans and biota, work as an integrated, interdependent system? What are the chains of cause and effect? Can we reduce what's happening to physics or mathematics, or approximate it in some quantitative way, if merely with arithmetic? Difficult as that may be when it comes to weather prediction, the task is far more complicated when assessments and outlooks for resources and ecosystems are at issue.[9] Only with a certain knowledge base and a level of understanding does it become useful to capture the data in a computer. And in order to have a true *forecast* as opposed to a *hindcast*—to be able to anticipate what the atmosphere and oceans and land surface will do beforehand versus explain only after the fact why the drought developed or the storm did the damage it did or the ecosystem collapsed—the computers must be sufficiently fast to run the needed calculations much more rapidly than the natural processes they represent are evolving. But the challenge doesn't end there. An error of a tenth of one percent (far smaller than the skill of even the most precise instruments) in any single one of these

---

9. The Intergovernmental Panel on Climate Change (IPCC) scenarios for future emissions show the difficulties here.

parameters can sometimes discombobulate the forecast outcome.[10] What happens when/if one, or a few, of the measurements input into a model is not just in error but totally wrong? Suppose the measurement is coming from a bad instrument or a sensor that's experienced a power failure. Such bad data can interfere with the computer's ability to digest the data stream in much the same way that a toxin or some *E. coli* in our food disrupts our digestion. To cope with these problems, computers spend much of their power on so-called *data assimilation*, which handles massive data streams while looking for outliers, using a mix of techniques such as comparing observations from one site with those from other nearby sensors and comparing the data with the expected conditions at the data-collection point based on earlier model runs.

Because there's virtually no limit to what we might need or want to measure and where, there's a considerable gap between that ideal and what we actually *do* measure. A number of factors contribute to this gap. First, there's the sheer technological difficulty. As result we often measure what's easy to measure as a surrogate for what we really want. For example, satellite radiometry measures atmospheric radiance at different wavelengths as a surrogate for atmospheric temperature and water vapor measurements. These physical parameters of interest are then inferred through a complicated mathematical analysis known as inversion. Limits to computing power limit the ability to digest and use the data available; meteorologists are forced to prioritize, to choose between increasing spatial and temporal resolution, say, or to choose between resolution and faithful rendering of microphysical processes, radiative transfer, and more.

5.4. A Marshall Plan for the 21st Century[11]

Given that we're likely underinvesting in the critical infrastructure known as Earth OSS, that we lack a clear analysis of the ROI, and that we're paying a high price for belatedly learning what we need to know about Earth and how it works, how might we approach the challenge?

Here's one suggestion. It's based on three premises:

1. *Mankind's defining 21st-century challenge is the threefold problem of resource acquisition (especially food, water, and energy), protection of the environment and critical ecosystem services, and building global resilience to*

---

10. As discussed in more detail in Chapter 9.

11. Adapted in part from a post on LivingontheRealWorld, dated January 9, 2013: "WWBD: What Would Boone Do?"

hazards at the community level. The tricky part is that these three goals must be achieved simultaneously and in such a way as to give the generations who follow us in the 22nd century a fighting chance to leave a similar legacy for those who follow them in the 23rd century and keep the human race prospering.

2. *Earth OSS is the starting point.* All the satellites, radars, surface probes, drilling and prospecting, computer modeling, and professionals doing good *thinking* are the essential foundation for this work. At something like $15 billion a year in the United States and maybe $70 billion a year worldwide, out of a $15 trillion a year in U.S. GDP and $70 trillion a year in global GDP, these tools are relatively inexpensive.

3. *Doubling (say) this annual investment in Earth OSS would only have to catalyze an increase in GDP growth rate (the global GDP growth rate is about 3.7 percent per year) of some 0.1 percent per year to pay for itself.*

Just a brief further word on each of these. The first is the so-called sustainable development challenge. Remember, a steady-state sustainable development is actually an oxymoron; the best we can do is buy time. This means a continuing investment in innovation, getting more efficient in every aspect of our lives, finding new substitutes for virtually everything we use, and so on. Second, Earth OSS is an infrastructure every bit as critical to our future as the electrical grid, or communications, or water, or transportation. Third, the doubling figure might seem to have been plucked out of the air, but it's actually the figure that makes the best sense. A 10 percent increment doesn't dramatically change our capabilities, and a factor of 10 increase is too great to assimilate effectively; we would just waste it. But the doubling represents such a small investment (in the United States, 25 percent of the supplemental budget request for Hurricane Sandy damages, for example) that we wouldn't need to wait for an extensive advance analysis of the anticipated economic benefit. We could instead afford to simply make the investment and then measure the resulting ROI over the first few years. If we liked the result, we could double the investment again. Third, writing in the *Bulletin of the AMS*, Jeff Lazo and his coauthors showed in 2011 that year-to-year weather variability in U.S. GDP amounts to about $500 billion per year. If we were to use better weather information to be 3 percent smarter in the way we capture weather's benefits and avoid its hazards, our increased investment would more than pay for itself. And that's saying nothing about the other benefits from such investments.

Domestic U.S. benefits are just the beginning. Consider this. The last time the United States was in a financial pickle comparable to the one we face today was the end of World War II. Today we have a GDP of $15 trillion per year and a national debt of $18 trillion. Back then we had a GDP of $150 billion per year (only 1 percent of today's!) and a national debt of $180 billion.

What did the United States do? Did it decide to hunker down, retrench, lick its wounds, focus on needs at home, do more with less, or put on that straitjacket?

No.

We (more correctly, our parents and grandparents) made the decision to spend $15 billion over four years *not in the United States*, but to rebuild the economy of Europe—the so-called Marshall Plan.

To emphasize the comparison, today that would be the equivalent of Congress deciding to spend $1.5 trillion spread over the next four years on the rest of the world.

Amazing.

Why did the postwar United States do that? To prevent the spread of Communism through Europe. And it worked like a charm. In fact, it didn't simply contain Communism. It worked so well that over the several years following the war, the U.S. GDP doubled, jumping to $300 billion per year, in part because by means of that postwar reconstruction we created a new trading partner.

The economic benefits continued. The rapid growth gave the United States the resources that 40 years later allowed President Reagan to force the collapse of the former Soviet Union and bring the Cold War to a close in the late 1980s.

What might be the goal of a corresponding investment today—a Marshall Plan for the 21st century? Well, it doesn't seem that the world's economies need such mending. And to prime the pump enough to make a difference would require staggering sums. But there is another powerful trend going on. The developing world has never been richer, in part because the developed world is coming in to their countries and locking up natural resources for both the present and the future. Two major examples: First, China is buying up minerals such as iron and copper worldwide but especially throughout Africa and across Southeast Asia. Second, both China and Saudi Arabia are buying up African arable land, ensuring food supplies for their people for future years. Although the wealth transfers are unprecedented, the sellers might be forgiven for wondering whether they're receiving enough in return for what they're giving away. In fact, Myanmar and other nations seem to be reaching out to the west these days for precisely this reason; outside international interest in their resources is making them skittish.

Suppose the United States (and any other countries willing to join), in the spirit of the Marshall Plan, were to partner with the developing world in helping them improve their understanding of what such assets were worth?

That would likely create as much good will toward the United States as the Marshall Plan generated 60 years ago.

The cost to the United States? Nothing more than a greater investment in Earth OSS. The figure wouldn't be close to $1.5 trillion either—more likely 1 percent to 5 percent of that each year, maintained for 10 years, say. Though the benefits would be international, much of the actual expenditure would be spent domestically, to purchase expertise and services from U.S. companies and professionals. And chances are good that other nations would want to join us and help.

Our United States can choose to be defeatist—to figure that times are tough and decline is inevitable—and in that way contribute to a self-fulfilling prophesy.

Or we can start to think and behave more like George Marshall's post–World War II generation and self-fulfill a prophesy of a more positive kind: fostering rapid economic growth, protecting the environment, contributing to geopolitical stability and national security, and building resilience to natural hazards. That rationale leads to the second proposal of this book (see Chapter 2).

Real-World Living Initiative #2. Double the rate of investment in Earth OSS in the United States (and hopefully, for that matter, worldwide). Reap the domestic benefits. But don't stop there. Develop and implement a Marshall Plan for the 21st century.

The physical nature of the real world makes a compelling case for such an initiative. But to achieve such a goal requires that we understand and appreciate the *social* realities that govern our world as well.

5.5. A Basis of (Social) Facts?

As Chapter 3 reminded us, just as it's important to understand how the geophysical world works, it's essential to comprehend how individuals, organizations, and societies function. As complex as the physical world is, the social world is even more so. Mathematics offers far less help to the social sciences than to the physical and natural sciences. The social sciences are arguably far more broad, spanning psychology and its subdisciplines, sociology, political science, history, economics, and much more. However, federal funding for the social sciences is 10 to 100 times smaller than that for the physical and natural sciences.

Several factors contribute to this difference. First, the work is less costly. Particle accelerators, large telescopes or radio-astronomy arrays, research ships and aircraft, and satellites are not involved. Second, in the years since World War II, physical scientists have played a leading role in making the budget allocations for research (more on that in the next chapter). Third, policymakers consider the physical sciences to be apolitical or nonpartisan, while much social research has a clear and close connection to politics. Consider, for example, studies of the influence of poverty on crime, health, or educational opportunity; the effects of legal and illegal immigration on demographics, voting patterns, and the economy; or the links between demographics and gun use or between religious views and abortion. To academics, such studies may look like basic research; to politicians, they look more like assault weapons in the hands of opponents. (Or put another way: If we can beat our swords into plowshares, then plowshares should not be welcomed but rather feared as potential weapons.)

More on this topic in the next chapters.

5.6. Recap

If we could first know where we are, and whither we are tending, we could then better judge what to do, and how to do it. —*Abraham Lincoln*[12]

If we're going to make a success of living on the real world, we're going to need a lot more knowledge about the way things *are* (the physical *and* the social realities) and a greater understanding about where events (and we) are headed. To say that making decisions and taking action without knowing is hugely costly is an understatement. Flying blind into the future is what the metaphor suggests: a prescription for repeated disaster and tragedy. Realizing belatedly the implications of our mistakes is hardly better, unless we or others learn from such experience. What's worse is that throughout earlier human experience, our smaller numbers and lighter environmental impact allowed us simply to correct mistakes. But today, with seven billion of us consuming resources, impacting the environment and ecosystems at a fast rate, and more exposed than ever to natural hazards, the bar has been set far higher. We need to build a level of knowledge and understanding that

---

12. Lincoln's opening statement in his famous "house-divided" speech, made on June 17, 1858, at the Illinois State Capitol, on the acceptance of his nomination to run for the U.S. Senate against Stephen A. Douglas.

allows us, like a master chess player, to see the future implications of our present options.

But while knowledge and understanding constitute an essential starting point, an important tool in the kit, they're by no means sufficient unto themselves. We need the wisdom of Lincoln. We need to *better judge what to do, and how to do it*. We need a tool for translating knowledge and understanding into effective decisions and action.

That's where policy comes in.

6

POLICY

Go to the ant, you sluggard: consider its ways, and be wise! —*Proverbs 6:6*

Over time, it seems that the word *policy* has been imbued with much more mystery and more majesty than it deserves (maybe in that respect it's similar to science). But policy is not a realm reserved for (over-) educated people, a privileged few at the top of some legislative heap, or at this or that ivory tower or think tank. The fact is, it's something far more important, and it's almost just the opposite. We might with equal accuracy view policy as the last desperate measure we all take when we find ourselves not smart or quick enough to make all the decisions we face any other way. It's our way of leveraging our intelligence: speeding up decisions and making them more effective at the same time.

Living on the real world is challenging. The resource, environmental, and hazard risk-management decisions all seven billion of us have to make are complicated, and they come at us fast, jumbled atop one another, mixed together, and interconnected. We don't have all the time in the world to make up our minds; we aren't given very many mulligans or do-overs. And, frankly, we're not bright or quick enough to make each new decision de novo, on its individual merits, starting from scratch, as it comes.

Two realities make our 21st-century task even more intimidating. First, even after all our history, we're only just beginning to understand how Earth and its ecosystems work. The basis of facts we'll be accumulating this century (see Chapter 5) will reveal bewildering complexity and convoluted intricacy. We may eventually *know*, but for much of this century we'll be struck by what we *don't know*.

Perhaps more daunting, in the Age of Interdependence we're all mutually connected. In most circumstances and for most issues the decisions and actions we need to make are affected by decisions and actions others are taking, in many cases nearby but increasingly a world away. In turn, what we decide and do at our end changes many situations abroad. We have little time or opportunity to consult the other people about what they've done or what they're contemplating doing next, or what they hope for or need from us. Instead, much as we rely on higher-level computer languages instead of programming in machine code, we rely on *policies* to speed up our decisions and actions; to make them more effective; and to provide continuity, stability, and quality control.

This chapter and the next two will examine policies and how they work in a bit more detail. We start with a few examples chosen to emphasize these points—to help us press the reset button on our own prior thinking with respect to policy.

6.1. Ants[1]

The biblical proverb of course encourages us to esteem ants for their industry, but a closer inspection shows that there's much more to admire. If you were to encounter an isolated ant, you'd find it moving around more or less aimlessly. Alone, it seems to have only limited wisdom. But aggregate a few of them together, and a social intelligence emerges—a property that you would never have suspected no matter how long you had watched that individual ant in isolation.

The ants collaborate to build a colony, topped by an anthill, which is marvelously, intricately complex. No two are alike and yet each turns out to be a perfect home. It's structurally sound. It's got HVAC to allow fresh air

---

1. I owe my introduction to the subject to a wonderful little collection of essays written by Lewis Thomas titled *The Lives of a Cell* (Viking Press, 1974). Of course, much original work in this and related areas was done by Edward O. Wilson, the myrmecologist and social biologist, and those who have followed him and built on his work.

in and control temperature and humidity so that the ants thrive. It features access for food to be brought in and egress for waste, and it protects against predators.

But there's no ant you can accuse of having this big picture. Instead, there's something in their very simple coding[2] that in effect tells them things like "we should start here and not there. We should start now and not later. This dirt particle should be taken off the roof of this tunnel and carried outside the anthill, and that dirt particle should absolutely remain where it is. This ant should be attacked, while that ant should be welcomed or at least accommodated. I am a worker and not a soldier. . . ."

Ants use related techniques—going under the name *stigmergy*—in acquiring food. A given search originating from a colony may start out randomly. But when an ant finds food, it returns to the colony . . . more or less (but not perfectly) directly . . . laying down a pheromone trail on the way. Other ants subsequently follow this trail, and they lay down pheromones on their return as well. But the pheromones evaporate. Thus if there are two such trails, the shorter picks up higher pheromone levels and attracts more ants. Again, the ants in this way work out an optimization. Termites use pheromones in similar ways in construction.

The ants are clearly *living in the moment* and *thinking inside the box*.

They're myopic and unquestioning. And yet they're so perfectly attuned to that moment and to their local circumstances that their actions work together and add up over substantial periods of time to enhance and maintain their long-term survival. They're meeting their resource needs, preserving their environment, and managing their risks *simultaneously*. In microcosm, they're coping successfully with the same challenge we face. In fact, the word *successful* hardly begins to cover it. The combined population of the 14,000 species of ants number in the quadrillions; their combined weight is comparable to that of humankind. They've been around perhaps 100 million years or so.

Hmm. It's easy to admire all this. But at first blush, this idea about living in the moment and thinking inside the box might seem wrongheaded. Our entire lives, starting (and continuing) with our parents, mentors have been encouraging us to go another way. They've impressed on us the importance of thinking about our long-range future instead of letting ourselves be captured by the temptation of the moment. Thinking long-range is supposed

---

2. Thanks to the work of Mr. Wilson and many others, we know that this is mostly chemical.

to encourage us to do our homework instead of goofing off with friends, to choose friends of character rather than the most popular, or to eat our vegetables. We're told thinking out of the box is the essential starting point for creativity and innovation.

All true.

But the idea here is that these valuable approaches to life work far better if we bring them to the point where they're instinctive instead of relearned each moment. As teenagers, we're better off if we refuse cigarettes, alcohol, empty carbs, and other addictive substances out of hand rather than consider the issue afresh with each new invitation from our schoolmates. And we can't delay local actions until some larger global framework can be put in place.[3]

To reiterate: No matter how smart we might be, the brightest among us are incapable of thinking through the right thing for humanity to do in the exciting but tumultuous, brawling, coming-at-you-fast 21st century. But if we can follow the advice of the proverb, perhaps figure out what an analogous strategy might look like for us, then perhaps we can successfully navigate our own future. Mathematicians and computer scientists are already modeling such ant behavior with an eye to solving human problems.

The bad news is that ants don't have an inflated view of their mental abilities, whereas we do. They're dumb and they embrace it. We're not that smart either, relative to the challenges we face, but we live in an unhealthy denial. Remember that contradictory certitude described in Chapter 2 that is a characteristic of wicked problems? Each of us, all of us, knows exactly what to do about the world's ills. Just ask! We'll give you an earful. About climate change. About disasters. About world hunger. About healthcare, jobs, or terrorists. About today's youth (especially if we're older). About today's elderly (especially if we're young).

You name it, we have an answer. The problem is that we don't agree. Our individual proposed fixes are conflicting. Probe in sufficient detail, and you'll find that like snowflakes no two answers are the same. Leave a conversation with any single one of us, and you'll know exactly what to do. Read any of today's best-selling books, and you're equipped to take on the world. But

---

3. For some, this may call to mind the notion of "think globally, act locally," which has been used in various contexts but is most often associated with environmental issues: acting locally to do things that if widely adopted would improve the health of the planet as a whole. The idea if not the exact phrase has often been attributed to Patrick Geddes (1854–1932), a Scottish city planner (and accomplished in other areas as well).

join in any larger conversation—or read a second book—and you'll walk away shaking your head.

There's more bad news. Ants have other attributes that we struggle to emulate:

Within the ant colony, the ants put the larger interests of the group ahead of their own. There's little evidence of envy, jealousy, or grumbling. There don't seem to be any slackers. No one's goofing off. When a would-be predator intrudes, those ants called to fight and sacrifice for the good of the larger group don't shy from this responsibility. They don't question why they and not others have to take the hit. There's little evidence of a NIMBY (not-in-my-backyard) syndrome. Ants also seem to be loyal. (A small aside: Whereas most ants confine their loyalty to those in the immediate anthill, Argentine ants have taken this a step further. Whatever genetic coding governs their behavior is such that they can be removed from their home colonies and put in another Argentine anthill half a world away and still be accepted into the group. Not surprisingly, this seems to confer survival value; Argentine ants, though small, devastate colonies of other ants when they come into contact. Here and there, wherever they've been accidentally introduced, they seem to be in the early stages of taking over.)

Those who've studied the sociality of insects have come up with a word for all this. They tell us that ants are *eusocial* (combining the Greek word *eu*, meaning good or real, and *social*). For ants, living on the real world means cooperative brood care, overlapping adult generations, and coexistence of reproductive and (partially) nonreproductive groups dividing the work. Bees, wasps, and termites exhibit similar behavior.

By contrast, apart from small groups, such as our family, military platoon, office colleagues, lab mates, or faith community, we humans seem to have trust issues. For most of us in most areas of our lives, our circle of interdependence extends far beyond our circle of trust. More broadly, we hold virtue in high esteem (there's a whiff of tautology in that statement) in the abstract, but we struggle in the execution.

The ants' policy success also ties in to realism. Such policies become far more powerful when they're not just idiosyncratic but adopted widely, across society as a whole—when you and I can count on those with whom we live and work and share the planet to follow that policy as well.

But before turning to public policies, let's look at one more example from the animal kingdom, a case for which scientists may have worked out what the policies actually are.

6.2. Boids

Sampling a few links on YouTube under the topics "birds swarming" and "boids" makes for some interesting and instructive viewing.

The first grouping offers strikingly beautiful videos of birds flocking by the thousands or tens of thousands, forming intricately organized, wondrously evolving patterns. The second grouping takes the viewer to computer simulations that render remarkably faithful, detailed imitations of the same thing, such as swarming in the urban environment, through its many city canyons and obstacles, and swarming in the presence of predators.

In each case, you might be prompted to ask: How on earth do birds do that, especially considering they're birdbrained? For that matter, how can computer programmers possibly simulate it? Well, it turns out we can thank Craig Reynolds[4] and others who did some of the first animations, going back to the 1980s. Mr. Reynolds programmed three so-called steering behaviors:

- separation: steer to avoid crowding nearby birds (avoid collisions);
- alignment: steer toward the average heading of nearby birds (match speed and direction); and
- cohesion: steer to move toward the average position (center of mass) of nearby birds (no single bird maintains the lead).

When he ran the simulations, they looked pretty much like the real thing.

We can't read each other's minds. And for that matter, we can't read a bird's brain. So we can never really know whether Reynolds has got it right. But we can infer that something like this hypothesized thought process must be going on. It allows each bird to make individual decisions, based on local conditions at the moment, rather than struggling to develop some more comprehensive view. Standing at some distance from the flock, we marvel at the aesthetics of it. But for birds imbedded in the flock (and for insects in the swarm and fish in the school), this behavior has practical value, most decidedly when all the birds do it, for example, in the presence of predators.

Human beings exhibit a similar, equally purposeful behavior.

For example, something like this same policy comes into play twice every weekday during my Metro commute here in Washington, D.C. At the height of rush hour, trains disgorge hundreds of passengers every few minutes at

---

4. Reynolds also came up with the term "boids" as in birdlike object, but conveniently evoking a New Yorker's stereotypical accent.

the dozens of stations throughout the region. Floods of people continuously stream smoothly past each other, changing trains and entering or leaving the system. Collisions and bumps happen (they happen in the computerized simulations as well), but they are statistically rare. Most people on most days avoid any such awkwardness. They merge into lines flowing in pretty much the same direction. No one has been assigned any kind of lead role. The entire show is self-organizing. All is well.

Except for tourists and visitors. They don't know the intricacies of the system: which exit best leads to their destination, which platform serves their train line, or which escalators have been out of commission for the last month. As they hesitate, stop to consult, huddle to work out a plan, they inconvenience others and pose risk to themselves. When we visit their home cities—Tokyo or London or Moscow or San Francisco—we find our situations reversed.

In fact, there's a difference between the behavior of visitors from another big city and visitors from small towns and rural areas. The former, when in a strange Metro system, know the business afoot and how to behave. If they have hesitation, they quickly step to one side. They find a quiet spot out of the main flow to ponder their next move. The latter will sometimes simply stop in their tracks. I found this out the hard way, moving from relatively rural Boulder, Colorado, to Washington, D.C. I quickly learned that I was at heart a far too uncooperative person. Driving in rush hour traffic proved to be a similar circumstance; I'd find myself in the wrong traffic lanes. Or I'd see a long line of cars building up and I'd cruise past, figuring I could cut back in to the lane I wanted up ahead a few hundred yards. Though car horns provide only the crudest communication medium, I quickly was forced to see that most of the rules at most intersections are unwritten and to start respecting my fellow commuters' code of behavior. I transitioned over time from part of the problem to part of the team. The rap on city people, going back to children's stories about the country mouse and the city mouse, is that the city dwellers are cold and calculating while their country counterparts are all heart. But it turns out that city folks are far more cooperative. In a city, unless millions of people move like a precision drill team, nobody gets to work on time in the morning or makes it home promptly; nobody gets the job done during the allotted eight hours. City folks are in it together and they know it.

This interdependence is fundamentally different from the independent lifestyle of farmers living and working on the largely separate farms of 200 years ago—the American stereotype we carry around in our heads. We can legitimately celebrate our historic political independence from England, but in no way can we declare any independence from each other here at home. We need each other.

Birds and their policies merit a few additional comments before we move on. First, not all birds exhibit swarming behavior. For example, raptors tend to operate alone and to be more spatially dispersed. To the extent their diet includes small birds, swarming is an obstacle for them to overcome versus a practice for them to follow.

Second, birds operate under many policies, not just one. They have policies for seeking and obtaining food, other policies for choosing habitat, and still other policies for nesting and building nests, and so on. These are quite distinct from species to species. The policies are somehow different for a robin, a woodpecker, a duck, and a cowbird (which lays its eggs in the nests of others). They have migration policies and more. In many cases, when it comes to these policies, we may not have worked out a conjectural set of rules that replicates the behavior we've observed, but it's got to be there, just the same.

Third, some bird policies operate at a much more basic level. They speak to the detailed moving of each wing and the tail that is allowing the bird to stay airborne at all. All that muscular control is subconscious. We humans are mostly operating in the same way. Our so-called autonomic nervous system controls breathing and heartbeat and myriad other operations vital to our continued existence, but it operates totally underneath our radar screen. Rarely are we conscious of these actions, and when we are, it usually means we're on the verge of trouble.

Fourth, no bird ever has to consult another. That would be a complete showstopper. And nothing is said about *destination*. It's all about the *journey*. The birds are obeying a common set of rules, but there's no bird in charge. There's no top-down, command-and-control paradigm (remember from Chapter 3 that no one in charge is likely to be a good thing in many circumstances).

Fifth, and remember we're making much of this for our larger human context of real-world living, the birds are obeying their three steering rules *simultaneously*. They don't fly for a bit obeying just two of the rules and then switch to another pair or maybe fly a while conforming to a single one. They're always satisfying all three of their steering rules. If instead birds decided to change or relax or eliminate one or more of the rules, then the emergent outcome might become something quite different. Suppose, for example, the birds don't try to maintain separation. Collisions would become much more of a factor as the birds converged on a single point. Suppose each bird tries to lead; suppose the birds *compete*; suppose instead of *matching* speeds they try to outfly neighboring birds; suppose instead of flying toward the center of the group they were to try to reach some greater height relative to the others. Then the observed behavior might look quite different. And the *utility* of that behavior—the contribution to species survival, say—might be compromised. The strongest birds at the top of the swarm would be more isolated, easy prey for predators, and preferentially removed from the flock. This is vital to our understanding.

Sixth, it's not totally obvious just from inspection of birds' three steering rules that they should lead to swarming behavior at all. We have to program them and run the simulations to see what develops as the number of birds/boids increases. After the fact, you and I may think it self-evident, but this is primarily because Reynolds and his colleagues have provided a compelling demonstration. But the rules *don't* say to "perform massive, aesthetically appealing patterns in the air that change form rapidly, thwarting predators and giving each individual bird some measure of protection." Instead, swarming is what those who study chaos and complexity theory call an *emergent* property of the steering behaviors. The property *appears*, as more birds join in.

In this and the next several chapters, we'll look at what's involved in carrying this analogy over to human behavior and using policies to modify and control our behavior, especially when it comes to real-world decisions and actions with respect to resources, the environment, and hazards. We have some idea of our end goal: the elusive but seductive Holy Grail of sustainability. But we know less about just which simple policy bits—if practiced by all seven billion of us daily, locally, individually, and in aggregate—might get us there. That's going to take a lot of teasing out, a lot of learning from experience.

We're now ready to move on to public policy.

6.3. Public Policy

The fundamental policy on which laws rest, especially policy not yet enunciated in specific rules. —*Dictionary.com definition*

Most of the time when we encounter the word *policy* it's shorthand for *public policy*, as usually defined in some way similar to that above. Economic policy. Foreign policy. Trade policy. Science policy. Education policy. Water policy. Energy policy. Agricultural policy. Environmental policy. The social safety net.

These policies, in aggregate, both reflect and determine what kind of nation we were, are, and will become—our history, our present, and our destiny. In some cases, they provide incentives for some behaviors and discourage others. In other instances, they may simply allow behaviors. In other arenas, they regulate, prohibit, or mandate and require. Because they're policies for the real world, not some fiction writer's fantasy, they work best when they're grounded in physical, social, and spiritual reality.

One of the great triumphs of our Founding Fathers, who constructed the U.S. Constitution and the Bill of Rights, is that they started from a realistic premise: that most men and women are prone as individuals and as a collective to favor short-term self-interest, emotion, and personal comfort.

When it comes to the United States, the best example of this grounding in reality is also the broadest and most pervasive. One of the great triumphs of our Founding Fathers, who constructed the U.S. Constitution and the Bill of Rights, is that they started from a realistic premise: that most men and women are prone as individuals and as a collective to favor short-term self-interest, emotion, and personal comfort. Accordingly, they built a framework for doing business that ensures that as we all act in our self-interest, we at the same time contribute in some measure to a greater good.

This social contract has served the United States well for over 200 years. In fact, it has served so well that it has benefited the world's peoples, not just Americans, as other nations have modeled their constitutions after ours. But the social contract is beginning to fray a bit around the edges. If we're not careful, it may well unravel. The new wrinkle, the one that makes all of today's problems more challenging, is not that we've grown any more selfish but rather that we've gotten cleverer about how to identify and to preserve our self-interest in the face of the self-interest of others and at others' expense. What's more, the richest and best educated among us have gotten relatively shrewder than those who are less well-off. Worse still, we've grown even more loathe to yield the tiniest bit of compromise in this arena. This is manifest in every great national debate today: jobs, the social safety net, immigration, foreign policy, and more.[5]

---

5. Recall Chapter 3's discussion of wicked problems, the dimension of redistributive impacts, and strongly entrenched interests.

(And that introduces a fundamental notion: policies and regulations, and the rule of law, are limited when it comes to compensating for what at their heart are *spiritual* problems. Take, for example, the Golden Rule that we're not in competition with one another but rather esteem one another, having some sense that we're all in it together. It's hard, indeed impossible, to legislate this. For ants, tens of millions of years of natural selection have weeded out the selfish behavior. We don't enjoy the luxury of so much time. The will to conform to such values has to come from some other source.)

As a way of illustrating the points we've developed above—that policies have a huge influence on outcomes and that the stakes for inaction, for assuming that policies are sacrosanct, are high—let's take a moment to examine two policy topics, rather removed from the main thrust of this book. They are themselves polarizing, but hopefully because they *are* slightly off topic they might allow us to remain emotionally detached, and therefore more likely to see the importance of the policy process to the human prospect.

1. *Drug policy.* In the United States, marijuana, cocaine, and heroin use are criminal offenses. The goal behind this policy is to keep our population free from the scourge of drug addiction, which ruins the lives of thousands, if not millions, of people in the United States alone. Every individual life wasted in the thrall of such addiction is a profound tragedy. But criminalization of these drugs has had the perverse effect of incentivizing their production and marketing, especially in the poorer nations to our south, and in the process creating a subculture of crime and violence throughout Central America, extending into South America and the United States as well. Here in the United States, it has contributed to our dubious distinction of having the largest per capita prison population in the developed world. The economic and social costs of this are enormous. And wholesale aerial fumigation of the fields where these crops are grown leads to environmental damage and public health consequences for those living nearby.

In this light, some have proposed decriminalizing such drugs. At the state level, some experiments are underway to examine the implications of alternate policies. Data from abroad (e.g., the Dutch) also provide useful background. Our own historic experience, with criminalization of alcohol during Prohibition (1920–1933), suggests that criminalization does drive up corruption and violence and that elimination of these laws causes a drop in prices, largely eliminating incentives for drug dealers at every link across the supply chain, without leading to much, if any, increase in addiction. Few voices are proposing a return to Prohibition any time soon.

2. *Immigration policy.* In brief: America is largely a nation of immigrants, as opposed to indigenous people.[6] The immigration policy here has a correspondingly rich history. It has varied over the life of the nation between expansionary, welcoming cycles and cycles where the borders have been relatively closed. Experts tell us that immigration has had two positive effects on the United States in recent years: (1) the U.S. population is aging more slowly than the populations of Europe and eastern Asia (China, Japan, Korea); and (2) U.S. science and technology owes much of its recent progress to contributions from foreign nationals who have come to the country for higher education, and have stayed. One counterargument is that immigrants, particularly illegal immigrants, take jobs away from American citizens, but even here some experts conclude that in many instances, new immigrants shoulder low-level jobs that most U.S.-born citizens don't want to do. As with drug policy, analysis suggests that immigration policy also holds consequences for resource availability and for the environment, as well as vulnerability to natural hazards.

In either of these two cases, we're not done with policy formulation, nor should we be. And we can easily imagine that changes in either or both of these policies, because they're intertwined, might profoundly change future U.S. prospects.

With this background, we're ready to consider the next policy

6.4. Public Policy for Science

Why is it that our federal government supports science? What's the basis? What's the history? It turns out that there's a constitutional basis, but it's remarkably terse. You can find it in Article 1, Section 8, which includes among congressional powers the power to "promote the progress of science and the useful arts," but this clause goes on to add, ". . . by securing for limited times to authors and inventors the exclusive rights to their respective writings and discoveries." Clearly this calls more for patent and copyright protection than any research support per se. Nevertheless, there it is.

There's a full and wonderful history of how this simple constitutional mandate has played out against the backdrop of all the great trends and events that shaped our American character: The continued effort to break

---

6. From what the archaeologists have told us, perhaps even our own so-called indigenous people might more accurately be described as the nonindigenous people who got here first, ca. 12,000 B.C., migrating from Asia through Siberia across a strip of land that has since become the Bering Strait.

away from Europe, manifested in the War of 1812; the expansion west, the great schism occasioned by slavery and its resolution in the Civil War; the Industrial Revolution; and the reengagement of the United States with the rest of the world, first through two world wars and more recently through trade. This history goes far beyond the scope of this book.[7]

But if we pick up this narrative at the end of World War II, we find this: The nation breathed a collective sigh of relief. There was awareness that we'd won the war and been spared some of its worst effects (those experienced by Europe and Japan) by dint of geography but also by accidental good fortune. As much by luck as by design, we were able to develop radar sufficiently early to blunt the might of the German Luftwaffe, which had been bludgeoning Great Britain. By good fortune we just barely beat the Germans to development of the atomic bomb. By a stroke of luck the war ended before Germany was able to bring its missile technology fully to bear on Allied targets. Serendipity gave us sulfa drugs in time to prevent greater impact of debilitating infections and disease on our troops in the South Pacific.

U.S. leaders took a look around at an emerging Soviet/Communist threat and other world trends and decided that we might be equally lucky the next time around, but perhaps we shouldn't chance it. Perhaps we should aggressively and strategically pursue scientific and technical innovation in order to maintain a strong position in world affairs and to protect cherished values of freedom and democracy.

So, under the leadership and vision of Vannevar Bush (you can find more on this in his landmark 1945 book, *Science: The Endless Frontier*, which is also available online) the United States established the National Science Foundation (NSF). The NSF was established in 1950 by public law 89-507, which frames the mission of NSF as follows: "To promote the progress of science; to advance the national health, prosperity, and welfare; to secure the national defense. . . ." Here the argument appears to blend Article 1, Section 8, with broader constitutional purposes.

This gets much closer to the heart of the issue, doesn't it? So let's take a fresh look at the Constitution's Preamble: "We the People of the United States, in Order to form a more perfect Union, establish Justice, insure domestic Tranquility, provide for the common defence, promote the general

---

7. If interested, you can find a thorough, imminently readable account by A. Hunter Dupree, *Science in the Federal Government: A History of Policies and Activities to 1940.* (1957) Harper Torchbooks (The book is out of print, but you can find the entire text online.)

Welfare, and secure the Blessings of Liberty to ourselves and our Posterity, do ordain and establish this Constitution for the United States of America."

Hmm.

Sounds familiar. That's because the NSF framers wanted to be sure to tie it back to the constitutional policy of promoting the useful arts. But they wanted to do more. In an era hot on the heels of the Great Depression and World War II, when the country was feeling beleaguered and poor, they wanted everyone to understand that science and technology was integral to basic U.S. values and purposes.

In the fast-paced, rapidly changing world of 2013, science looks not just helpful but essential to these aims.

Promote the general welfare? Try growing the economy, providing fulfilling, high-paying jobs, and protecting the environment and ecosystems any other way. We're not going to accomplish this without science and engineering to create and refine new products and services and keep us competitive in the world marketplace, or without adding to our knowledge and understanding how biodiversity, ecosystems, and humanity interact.

Promote common defense and secure the blessings of liberty? Our small population (300 million) cannot maintain a strong international leadership role in a much larger world of seven billion people without the added leverage that an edge in science and technology provides. More importantly, as every leader of our armed forces reminds us, military conflict should only be a last resort. Physical and social science and technology are vital to creating the economic and environmental conditions worldwide needed for geopolitical stability—to head off conflict versus merely win it.[8] And how to accomplish all this not only for the present generation but also for posterity—to be sustainable? Science holds the keys.

NSF funds much but not all of the basic research in the United States. What about the constitutional justification for the remainder? A little more investigation shows similar rationale for the basic research done at the National Institutes of Health (NIH), at the National Aeronautics and Space Administration (NASA), and across the Department of Defense (DoD). In addition, the general national aspirations captured in the preamble also justify the mission-oriented research of agencies including but not limited to the U.S. Department of Agriculture (USDA), the National Oceanic and Atmospheric Administration (NOAA), the National Institute of Standards

---

8. A Marshall Plan for the 21st century, as described in Chapter 5, comes to mind.

and Technology (NIST), the Environmental Protection Agency (EPA), the Department of Education (DoE), the U.S. Geological Survey (USGS) and other interior agencies, the Department of Human Services (DHS), and so on. Public safety in the face of hazards; adequate food supply; secure and ample water resources; clean, cheap energy; breathable air: All these goals and many more require an ever-stronger foundation of knowledge and understanding.

Now let's drill down a little more. Our political leaders and our scientist leaders at the time could have chosen to implement that intent in a number of ways. For example, they could have established NSF as a substantial bureaucracy, with in-house research capability and research programs, like the Manhattan Project writ large, and made it sustainable. (In fact, the nation today has many in-house federal research laboratories, building the basic knowledge and understanding of every subject under the sun.)

Instead, they went in another direction. They made it a policy that NSF would accomplish its ends through external grants, funding research in the nation's universities, spreading the support over a large range of disciplines, and funding research in those disciplines on the basis of proposals against those funds and extensive peer review.

Other agencies—the mission agencies—made different choices. The Office of Naval Research chose to emphasize more extensive and comprehensive support of institutions such as the University of Rhode Island, the Woods Hole Oceanographic Institution, and the Scripps Institution of Oceanography. The Department of Energy built on its support of Los Alamos, Oak Ridge, and other facilities during the Manhattan Project to establish a network of national laboratories today run by universities and other contractors. The USDA runs a mix of in-house laboratories and university-run projects based on Ag Extension Services dating all the way back to the Morrill Act of 1862. NOAA today runs a network of laboratories of diverse character, reflecting their various origins as part of the individual agencies that were merged to form NOAA, as well as a Sea Grant Network modeled after the Ag Extension Service.

Other countries chose alternative policy approaches as well. For example, the former Soviet Union, instead of supporting science and technology development through a diverse set of ministries operating on different models, chose a more monolithic, top-down approach. The result was to create something of a straitjacket for scientists and engineers. With only one path to the top, favoritism and cronyism and links to and conformance with the

Communist Party played a large role in the allocation of funds, promotions, and the like. This has combined with other flaws in Soviet society to inhibit the progress of Soviet science.[9]

Here's a remarkable feature of all this. For much of the past 50 to 60 years, U.S. scientists have operated under an extraordinary social contract, one that explicitly states that science has the best chance of supporting (very specific, very practical) national goals and priorities when it is unfettered, when it is motivated by bench-level-scientist perceptions of opportunity as opposed to top-down, command-and-control approaches or any temptation to try to pick winners in advance. It's as if scientists said to the nation "Give us lots of money, let us police ourselves and don't ask too many questions, and one day you'll be glad you did." And it's as if the country overwhelmingly agreed to this. Astonishingly, both sides have made good on this bargain for half a century.

But even though this social contract was forged during the Cold War, a period in history that seemed challenging at the time, the stresses on the United States didn't begin to approach those of today. Two new challenges have emerged. First, despite the rapid advance of science and technology, it appears that societal uptake has been fragmented and slow. And second, though science had once been considered nonpartisan, innovation in the political world has built an appreciation for how science might be used as a political weapon in debate on subjects as varied as climate change, stem cell research, evolution, genetic modification of plants and animals, and even particle physics.

Political leaders and scientists alike should find this sobering, for reasons that are in part similar and at the same time somewhat different.

Political leaders recognize that in the current atmosphere of fiscal exigency and partisan wrangling, support for science will pose an ongoing challenge rather a short-term problem readily resolved. It will require the best of their character, courage, responsibility, integrity, and, yes, their good nature to continue and augment science, to provide needed oversight, without smothering the needed creativity and innovation.

---

9. This should serve as a cautionary tale to those who would stamp out any science that looks the least bit duplicative across individuals and institutions. Such redundancy accelerates the detection of errors, and it provides multiple employment alternatives for scientists instead of making them vulnerable to the idiosyncrasies of a handful of leaders and thus fosters the advance of knowledge and understanding.

Scientists, for their part, will be asked to be strong, brave, responsible, and good-willed. They must keep in the forefront of their thinking that free rein for science does not mean freedom from responsibility. Public support for basic science or mission-related science is not an entitlement. The goal is still the resulting benefit to the American people (and the world's peoples by extension of the phrases relating to domestic tranquility, defense, and liberty; Americans cannot achieve these aims in isolation). Scientists would do well to consider frequently and with some degree of self-discipline the potential benefits (or possible risks) that might result from their work, and how to accelerate that benefit.

To date, both sides of the social contract have reason to be pleased with the results. American science is considered second to none worldwide. Advances in high-energy and condensed-matter physics—in the science of materials of every description, in nanotechnology, in high-performance computing, in the life sciences, and in the Earth sciences—have accelerated over the past half-century, created wholly new industries and opportunities, and generated unprecedented wealth. Students and scholars come from all over the world for the opportunity to study and do research in the United States, and many stay.

Over time, the NSF policy has morphed a bit. For example, today all proposals have to include, in addition to material relating to the intellectual merits of the proposed work, explicit material on the broader impacts anticipated to result from the proposed work. As a result of this policy change, the relevance of NSF research to national health, prosperity, and welfare, as well as national defense, is more clearly evident. Time will tell whether this policy change will accelerate the pace of scientific advance and associated societal benefit or slow the pace.

And that's precisely the point. The challenge is to conduct many more policy experiments such as this, learn from each one, and make corrections and adjustments over time in response to the results as they come in. And this learning needs to occur across the entire spectrum of policies, not just science policy. Currently, however, our approach is almost the exact opposite. The political debates around policy formulation have become so bitter and rancorous, and the emphasis so much on minority techniques for obstructing the process, that policy formulation (instead of occurring in more or less continuous, incremental, easily reversible steps across the board) occurs in massive spasms. Just a few examples include the healthcare debate throughout 2009–2010; the fiscal cliff of late 2012 to early 2013; or the sequestration and the furloughs of thousands of government workers, including government scientists. Call to mind the travel restrictions that prevent government scientists from engaging their peers and in the process being challenged to improve their science.

6.5. Science for Public Policy

> Get your facts first, and then you can distort them as much as you please.
> —*Mark Twain*

A casual look might leave the impression that public policy for science is no different from public policy toward foreign relations, or the economy, or healthcare, or education, or any of the national agenda. But that's not true. Closer inspection reveals a fundamental distinction. Unlike any of these other issues, science is able to *inform* policy, and advances in science hold the potential to improve the formulation of policy, not narrowly but across a broad spectrum. As a result, our choices with respect to public policy toward science work in special ways to shape the future of the United States and, for that matter, the global human prospect. If the United States aspires to be the essential nation for the 21st century, we must master public policy for science and the use of science for public policy. It's another of the several problems we have to solve *simultaneously*.

As mentioned earlier, public policy works best when it's based on and congruent with reality, both physical and social realities of the type we've been discussing. Let's consider an example, one of many. Compare, for the moment, monarchies and dictatorships on the one hand with democracy on the other. Dictatorships can and have worked. Occasionally (China constitutes a recent example) people tolerate authoritarian forms of government because they offer stability and short-term economic advantage. But totalitarian governments rely on force to maintain order, since people have a sovereign preference for individual freedom and representative democracy. That's the social reality. At the same time, democracy places demands on public education. Members of society need tools to hold their elected leaders accountable. That's another social reality.

Natural and social scientists, as scouts exploring and characterizing this reality (much as Lewis and Clark scouted out a young nation's Louisiana Purchase) therefore have much to offer the policy process. Significantly, this capability is relatively new. Governments, whether totalitarian or democratic, or somewhere in between, are still learning how to make fullest use of existing scientific understanding about the physical and the social nature of the real world and how to go about getting more of it. Examination of the U.S. Constitution shows only one reference to a federal role in obtaining and using data, and that is in Article 1, Section 2, which speaks to apportionment in the House of Representatives:

> Representatives and direct Taxes shall be apportioned among the several States which may be included within this Union, according to their respective Numbers, which shall be determined by adding to the whole Number of free Persons, including those bound to Service for a Term of Years, and excluding Indians not taxed, three fifths of all other Persons. The actual Enumeration shall be made within three Years after the first Meeting of the Congress of the United States, and within every subsequent Term of ten Years, in such Manner as they shall by Law direct. The Number of Representatives shall not exceed one for every thirty Thousand, but each State shall have at Least one Representative; and until such enumeration shall be made, the State of New Hampshire shall be entitled to chuse [sic] three, Massachusetts eight, Rhode-Island and Providence Plantations one, Connecticut five, New-York six, New Jersey four, Pennsylvania eight, Delaware one, Maryland six, Virginia ten, North Carolina five, South Carolina five, and Georgia three.[10]

The Constitution in this way led to the establishment of the Census Bureau.[11] There is no other mention of the use of science for policy: data on river flows and water resources as a basis for establishing and maintaining water rights; data on air quality or water quality as a guide to regulation of these matters; assessment of mineral resources, as a tool for allocating access to these; or developing agricultural policies based on crop production, etc. All that has been added since.

Such use of Earth observations, science, and services (the basis of facts described in Chapter 5, Earth OSS) for policy should rivet our attention.

Because this history is so recent, ways and means of using natural and social science in the policy process are still in their infancy. But one notion

---

10. Careful readers will quickly note that this language from the Constitution's original version tacitly or explicitly treats gender, indigenous people, and slaves in a manner that has since been repeatedly amended over the years in order to provide freedom and representation for all.

11. Ironically, given that social science has been considered the stepchild of policy for science, it was this social science (demography) that is the first use of science for policy in the United States.

is already clear. There was an initial phase very similar to the famous Garden of Eden before the fall. Data, and information and knowledge developed from such data, were initially welcome additions to the policy discussion. All parties understood science and a basis of facts to be helpful in constraining the dialogue and keeping the debate and consideration of options realistic. However, those Halcyon days are long gone. Policy officials, stakeholders, and scientists themselves have all more or less simultaneously awakened to the potential of data and science for manipulating debate and discussion to desired ends—so that science, instead of a common tool for creating a venue for civil discussion, has become a cudgel, a weapon in policy warfare.[12]

6.6. Recap

Policies are powerful tools, enabling societies to speed up decision-making, and translate knowledge and understanding into effective action. They're just what we need for converting scientific advance into societal benefit. At their best, policies help us leverage our individual intelligence into a far greater collective impact and influence. If they can do this much for birdbrains and ants, imagine how powerful they'd be in our hands!

However, ants and birds have had time to bring their (far simpler) policies to a higher stage of development. Millions of years of survival of the fittest have scrubbed the policies and sanded off their slightest imperfections. By contrast, the relevant policies here—governing resource development and use; environmental protection and hazards—are relatively new on the human landscape, and their utility has been compromised by rapid social change.

What's more, policies leverage our mental capacities precisely because their consequences are emergent. The same quality that gives them their power as we wield them either singly or in combination, on a large scale or spanning regions and nations or even globally also obscures their true implications and consequences at the beginning. We can usually fully comprehend their impacts, appreciate their finer virtues, and recognize their limitations only after they've been implemented on a large scale for extended periods of time.

Our collective task, therefore, has several steps: (1) to acquire a basis of facts: that is, knowledge and understanding about how the real world

---

12. A fair amount has been written on this. For an excellent starting point, you might try Roger Pielke Jr.'s book *The Honest Broker* (Cambridge University Press, 2007) and the references therein.

works, as discussed in Chapter 5; (2) to use these to formulate new policies and/or improve upon existing ones, as considered in Chapters 7 and 8; (3) to test our policy ideas: that is, try them on a small scale and quickly detect the emergent consequences of our policies (a suggestion for how to do this is offered in Chapter 9); and then, (4) should they prove effective, to widen their purview and use.

Chapter 7 quickly runs through different aspects of policies that might merit additional analysis and experimentation. The goal is to show that there's a great deal of work to be done, fairly urgently, and through such an overview to stimulate readers to join others in such policy analysis and experimentation across a broad front. Chapter 8 delves into one aspect of policy—natural hazards—to illustrate how in one particular instance it's possible to identify opportunities for improvement that would be inexpensive to implement and yet hold potential for significant improvements in outcomes. By extrapolation, we can expect to find that similar opportunities exist with respect to each and all of the other policy issues discussed in Chapter 7. But identifying candidate policies for improved real-world living is only the start of the job. For policies to be useful, they must exist as more than abstractions. They must be put into practice. Chapter 9 provides a brief discussion of how this works at the national level; it lays out a parallel with numerical weather prediction to suggest, in broad outline, how the policy frameworks under which societies function might be allowed or encouraged to evolve.

7

OPPORTUNITIES FOR IMPROVING POLICY

Had I been present at the creation, I would have given some useful hints for the better ordering of the universe. —*Alfonso X "the Wise," 1221–1284, Spanish king of Castile and Leon*

Few of us can match Alfonso's bravado, but all of us are going down a parallel path. We're constantly seeking to change the way we do things, and in so doing improve the prospects for our future world. Policy is one of a handful of tools that can get us where we need to be—inexpensively and on time. But if we are to succeed, it's not enough that policies help us make *faster* decisions. Policies must at the same time make those decisions *better*. We should constantly be in the business of examining, analyzing, modifying, and reinventing policies of every sort that would help us better comprehend and manage our world, or at least our part of it.

This chapter provides a very high-level[1] overview of some of the opportunities for improvement, incremental and revolutionary, that might make a difference. Each of the topics below could easily be expanded into a book or

---

1. The overview is also admittedly personal. Those interested should feel empowered to develop their own lists.

occupy a university research center for many years.[2] The hope and expectation is that in a population of seven billion, there are many who are already laboring along these lines, who have themselves identified these problems/opportunities, and who have made addressing them their life's work.

One obvious starting point? The three main threads where policies are required:

1. resources;
2. environment; and
3. hazards.

Each of these issues has an extensive body of policy. Hundreds of books and thousands of papers have been written on these three policy frameworks, considered as separate entities. Take resources. You and I can find entire literatures on water policy, agricultural policy, energy policy, and much more. Every nation and all languages offer their own source material. Furthermore, these literatures are themselves fractals. That is, look within water-resource policy, and you'll find policy on riverine water; on managed watersheds featuring dams, levees, and reservoirs; on groundwater; and so on. There's international, national, state, and local policy at the respective levels and sublevels. Look within riverine water policy, and you'll find policy on water quality, water amount, water pricing, and the role of the private sector in the ownership and provision of water, for instance. All of these policies, even those that are mature, can stand reworking and refinement in light of the current pace of scientific advance and social change. Water policy is a prime example; transboundary watersheds, serving multiple nations (or within the United States, multiple states), pose a particular challenge.

7.1. Synthesis

There exist extensive separate policy literatures with respect to resources, the environment, and hazards in isolation. Some work has been done to integrate resource and environmental issues. What's lacking is much in the way of policies that truly attempt to integrate all three as a coherent and consistent whole. The difficulties are substantial, perhaps prohibitively so, but

---

2. Some might wish for, or even have a right to expect, extensive references to such work here. My apologies.

mankind should be making the effort, much the same as we continue to look at the prospect for human exploration of Mars and other aspirational goals.

The motivation here carries back to the idea that humanity has to solve the three challenges *simultaneously*. At several points in the book, I've provided different ways of looking at this problem. Here's another: a simple set of three equations in three unknowns. Older readers might remember similar examples from junior high or middle school algebra; younger readers probably saw these problems in grade school:

> *The motivation here carries back to the idea that humanity has to solve the three challenges* simultaneously.

1. $x + y = 30$
2. $x + 2y - z = 20$
3. $3x - 2y + 2z = 50$

These have the unique solution $x = 10$; $y = 20$; $z = 30$.

Now suppose instead we just looked at the first equation in isolation. It would have several solutions, including but not limited to x = 10; y = 20. For example, $x = 20$; $y = 10$ would work. And so would $x = 2$; $y = 28$. And $x = 21$; $y = 9$. (If we allowed fractional answers, or negative integers, the number of possibilities would be infinite.) In the same way, when we focus our policy analysis just on resource acquisition, ignoring any need for environmental protection or resilience to hazards, it might seem that there are many solutions as opposed to just one. But as soon as we introduce the fuller set of considerations, the possibilities are constrained.

Note that in this example, if we were a bit in error about the right-hand side of the equations—say we'd measured those and come up with 30.05, 20.1, and 50.5, respectively—or the coefficients for x and y and z ($1.07x$ + $.95y = 30.05$, perhaps) our solution to the threefold set of equations wouldn't be much different from the original. *This suggests that a ham-fisted, clumsy approach to simultaneously dealing with resources, environment, and hazards might have more to recommend it than the most refined policy analysis of the three issues separately.*

Don't be misled by the brevity of this section. Such integrated policy should be sought and formulated as a matter of high priority.

7.2. Keeping Score

This will be discussed with respect to the hazards context in the next chapter. But policies with respect to resources and the environment could also be improved in this sense. We are inadequately applying full-cost accounting to our energy development, agricultural production, and water-resource management and allocation. By failing to internalize externalities and to eliminate subsidies we are postponing the day that our policies will be congruent with nature's realities.

By failing to internalize externalities and to eliminate subsidies we are postponing the day that our policies will be congruent with nature's realities.

7.3. Conflicts of Interest

As living on the real world has grown more complex, the range of interests at stake in any particular policy piece has grown. And with the proliferation of interests has come conflicts of interest of every sort and description. These should be managed—not just swept under the rug—and not merely tolerated. We need an improved ability to see these coming and a framework for dealing with them.

Two arenas for conflict of interest are of special concern here.

1. *Public-private-sector collaboration.* As discussed with respect to hazards policy in Chapter 8, the history of U.S. public policy has been one of maintaining an arm's-length separation between the public and private sectors. This has served the country well. Together with free-market principles and a set of cultural values they have minimized the corruption and graft that run rampant in many countries abroad. But when it comes to hazards policy and other issues where public and private interests are intertwined, the arm's-length policies are a straitjacket that inhibits broad national goals such as job creation and economic development, public health, and national security. We want to do better, but thus far policy solutions elude us.

2. *Development of knowledge and understanding versus use of that knowledge for policy formulation.* As indicated in Chapter 6, the U.S. Constitution is silent on the subject of data gathering, with the exception of the Census Bureau, needed for apportioning the members of the House of Representatives in Congress. In that one instance, there is a clear separation between the *gathering* of the data and knowledge (Census is housed within the U.S. Department of Commerce) and the use of those data (by Congress and by

the 50 states). For example, if state legislatures want to use Census data to gerrymander their Congressional districts, that's not the direct concern of the Census.[3] But for many scientific issues, and many issues relating to living on the real world, there's no such separation. And time and again, we find that *there is a conflict of interest between data collection and research on an issue and the formulation of policy based on that issue.*

Acid precipitation provides a case in point. We saw in Chapter 5 how the policy arguments led to a substantial, multiyear research program: NAPAP. But it was interesting to see the process up close, as I did from 1987 to 1991. Data gathered and research conducted by the Environmental Protection Agency (EPA) tended to show that acid precipitation was a substantial problem requiring environmental regulation (by coincidence, the EPA mission). By contrast, data collection and research conducted by the Department of Energy, loosely speaking charged with keeping electrical power reliable and affordable, tended to find much less of a problem. The findings of the National Oceanic and Atmospheric Administration (NOAA), whose mission was the measurement and prediction of the atmosphere, not surprisingly were usually somewhere in between. Canadian data, coming from a lightly populated country downwind of the more heavily populated and industrialized United States, tended to show more of a problem than corresponding U.S. findings. The same held true for studies conducted by environmental nongovernmental organizations (NGOs) on the one side and by coal companies and utilities on the other. Policymakers were stuck with the task of finding the middle ground between widely disparate research conclusions from all parties.[4]

This brings to mind the interesting history behind the U.S. Department of Commerce. When it was established in 1903 as the Department of Commerce and Labor by President Theodore Roosevelt (in 1907, labor was hived off to become a separate cabinet-level department), it was intended to be the

---

3. However, evidence is accumulating to teach us that such gerrymandering of Congressional districts is becoming more widespread and is creating "safe" districts for one or the other of the two major parties. As political offices in these districts are not in play across parties, the real political action is now occurring in the respective primaries. This fosters the success of extremist candidacies and contributes much to today's political polarization and gridlock.

4. Meanwhile, a similar problem was playing out with respect to climate change. Every once in a while, it would feel as if NASA findings were saying, "Earth is warming . . . and if our home planet proves too hot, we can move you to a cooler one."

statistical agency for the United States, charged with compiling all statistics with the exception of those pertaining to agriculture (by that time the USDA had already become an entrenched Washington agency). The fact is, if there's a single defining intellectual thread running across the entire Department of Commerce, it's the ability to measure things that are difficult to count or quantify. Some examples include global average temperatures; population of the homeless; fish stocks; the speed of light; the duration of a second of time; U.S. GDP and balance of payments; jobless rates; and much, much more. This history seems to have been lost over the intervening years. The Department of Commerce could do much more for the nation with respect to providing a neutral home for data collection and scientific evaluation than it has thus far been asked to do.

7.4. Triage[5]

You might be familiar with the term. It comes from the medical world. In wartime, or following a disaster, a hospital's medical capabilities can be swamped by the need. How to ration care?

Triage is one strategy. Picture a doctor or nurse giving each person coming in a first look as they are brought in for and await treatment, and then quickly assigning them to one of three categories: will not survive, can wait indefinitely, can or might survive but only if receiving immediate attention. There are fine points to this scheme, and variations, but you get the idea.

Whether or not you know or use the term, chances are good that you do triage yourself. Take e-mail. Hundreds of messages are coming in every day. When you check it after any kind of break, there's a new bunch. You scan the titles and senders quickly. Some you're never going to answer. How'd they get through your spam filter? Sooner or later, these will be deleted, but there's no urgency. They're goners! Some will get a response eventually, but they can also wait. Others you know you have to attend to immediately—and you do.

In the future, we're going to be increasingly using such an approach to environmental concerns. It's only natural to want to preserve everything, but the reality is that we have neither all the resources nor all the time in the world. Some species are probably doomed to extinction despite the most heroic measures we might contemplate. Some habitats (e.g., Superfund sites) have probably been so severely degraded as to be almost unusable going forward. Some ecosystems have been so profoundly disturbed that successive

---

5. Largely excerpted from a post by this title on LivingontheRealWorld, dated October 30, 2010.

states may be headed in an entirely new direction (think old-growth forests that have been logged or the regrowth around Mt. St. Helens following the 1980 volcanic eruption).

Here's just one real-life example, one I remember from my NOAA days. Several species of salmon were on the endangered species list at the time (being cautious here; they probably still are). Salmon are anadromous; that is, they spend most of their lives in the open ocean but swim up into freshwater rivers to spawn. These particular species spent the freshwater piece of their life cycles in the Pacific Northwest. Over the years, many of these watersheds had been dammed. The damming served multiple purposes: flood control; providing water for irrigation; hydropower, etc. Unfortunately the damming had a side consequence, making life difficult for the salmon, in several respects. First, the dams impose an obstacle to the newly hatched fingerlings trying to make their way to the sea. Some fingerlings, but relatively few, are killed going through the hydropower turbines. But that isn't the real threat. Prior to the dams, strong river currents would carry the fingerlings from the headwaters to the sea in just a few days. This kept loss to predation at a minimum. Since the damming, vulnerable fingerlings now spend large amounts of time being carried along by the slow currents in manmade reservoirs behind the dams. The journey to sea can take weeks. Loss to predation is now heavy. The return trip for the mature salmon is no picnic either. Swimming up the cascading rapids (the mountain range came by its name honestly) had been bad enough. But now the dams constitute an insurmountable obstacle. Bright idea: fish ladders running alongside the dam, replicating to some extent those natural cascades. But they're not a total solution. Predators, human and animal, have figured out the game and hang out by the fish ladders, waiting for dinner to be served. So, several salmon species are today endangered. The policy response has included trucking the fingerlings from headwater hatcheries to the ocean. Complicated. Labor-intensive. Expensive. And, despite all the cost and effort, only partially effective.

An alternative policy formulation has been proposed. It involves giving up on the mixed-use concept. Some watersheds would remain dammed. These watersheds would be devoted solely to electrical power generation, flood control, and water-resource management, with no regard for the preservation of the prevalent salmon species (such is the sophistication of taxonomy that analysis can often find distinguishing differences among fish from the individual watersheds). By contrast, the dams would be removed in other watersheds. Here salmon would be allowed free run; the other uses would be sacrificed.

Several additional points are worth noting. First, how much you like this policy approach depends on your culture, your values, and your ideals. If you favor hard-nosed considerations of costs (including opportunity costs) and effectiveness, you might reach one conclusion. If I favor tradition and culture and my tradition is farming and ranching, I might favor another. And if I favor tradition but I'm Tlingit or Chinook, I might favor a different approach still. My enthusiasm or agitation also depends on the policies prevailing on my particular watershed.

Second, in a given context, "good" or "bad" depends not just on the concept but also on the execution. Think back to the medical triage. How well trained and practiced are the doctors and nurses who are making these life-and-death decisions? What is their thought process? Is it cool and deliberative? Is it quick, almost instinctive, based on prior experience and practice? Remember, they're weighing, on the fly, much more than the extent of any injuries: the fitness level of the injured; their emotional state; the availability of further medical help behind that front door. What capabilities and staff are at hand? What kinds of injury, and how many people, is the ER equipped to handle? What would require further evacuation to, say, a burn unit at another hospital? These are gray areas. How are the doctors and nurses handling the uncertainties? How are they themselves holding up psychologically given the gravity of the situation?

Third, the medical world approaches triage with a considerable body of experience and fact. They teach to this experience. Triage is practiced daily in ERs across the country. By contrast, circumstances calling for environmental triage, while unfolding more slowly, are going to be far more complex. In many instances, they will be unprecedented. The basis of knowledge and understanding necessary for effective triage in most cases is not at hand. Nor will it be easily gained. We need far more investment in all research, including not just the natural sciences but also the social sciences, including but not limited to ecology and environmental economics, if we are to be well positioned for the environmental triage of the future. We also need to do some trial-and-error learning.

Fourth, you and I—all seven billion of us—will be the nurses and doctors in this scenario. Not just for this one small example of the salmon, but for the countless millions of similar decisions that are coming at us fast. We don't have much alternative. Geography won't provide an escape. Many of these decisions will have to be made right where you live. Your profession won't give you a pass either. You say you teach English literature? You're a journeyman plumber? A patent attorney? Drive a taxi? Already retired? It doesn't matter. You won't be able to excuse yourself. Chances are that many of these decisions will engage your profession directly. You'll be contributing to the perspectives and analysis that inform the discussion. And even if you aren't, as a member of the voting public in a democracy you'll be asked to help choose a course of action. Realistically, our futures are going to be full of these decisions. You and I all have to bone up.

Finally, to go back to its medical roots, triage is not so much the ideal as it is a strategy for making the best of horrific circumstances. As a future of triage looms on the horizon, we need to remember this: The aim is not just to master triage of greater scope and urgency, although that will surely need to be one goal; rather, the object is to master those decisions and actions that will keep the environmental triage needed to a minimum.

7.5. More Effectively Deal with Boundary Issues

Policy analysis and formulation would be difficult enough in a hypothetical world with no boundaries. But in the real world, boundaries of every type proliferate. The public-private-sector boundary problem has been discussed. But equally problematic for policies of all types are the boundaries separating government itself at all levels. Domestically, the local-, state-, and federal-level boundaries all pose problems for emergency response; for energy, agricultural, and water-resource management; and for environmental protection. International boundaries impose additional complexities. The boundary between natural and social sciences has proved problematic, as have the boundaries separating science of both these types from real-world practice. NGOs and civil society pose an additional challenge. The flip side is that these boundaries offer opportunities as well. One that we'll return to in Chapter 9 is the chance they provide to experiment with and test policies. Strikingly, for example, a number of U.S. states have instituted bits and pieces of carbon policy. California often provides what in effect is a pilot program for national policy on pollutant and emissions levels.

7.6. Inclusiveness[6]

These days everyone from our political leaders on down pays lip service to inclusiveness. Well and good! Some observations follow.

First, we sometimes make the mistake of thinking that societies and policymaking for those societies should be inclusive because we feel sorry

---

6. Much of this material came from posts on LivingontheRealWorld, dated December 8, 2010, and December 11, 2010.

for those who are often excluded. We view inclusiveness as a paternalistic responsibility—an act of noblesse oblige. It's just something we should do as a matter of etiquette, not something that we genuinely expect to be helpful. When we feel superior to another in this way and fail to respect the other's views and only make an external show of bringing everyone in, we're merely fostering hypocrisy versus seizing the opportunity that inclusiveness represents. In the 1960s, when the United States deliberately introduced policies of inclusiveness—first equal opportunity and then affirmative action—it was clear from the discussion, both pro and con, that many Americans thought this way. Interestingly, as artificial and as forced as these efforts sometimes were in their implementation, they've arguably strengthened us as a nation and improved our prospects, perhaps significantly so.

Consider for a moment our national and worldwide challenges: economic growth and reduction of poverty, public safety in the face of hazards, protection of the environment and ecosystems, education, and healthcare. Now reflect on our historic and present tendency to divide the country (or the world) into two groups, and then shut out one of those two groups in the problem-solving effort: men and women, elderly and young, white and people of color, affluent and poor, educated and not so educated, citizen and immigrant, people of faith and the nonreligious, public sector and private sector, Republican and Democrat. Why on earth should we want to do that? We've just halved our chances of getting the one bright idea that might solve the problem.[7]

Continuing in this same vein, one of the most interesting challenges for inclusiveness comes from today's social sciences. The majority of social-science studies involves dividing the world into two groups: investigators and subjects. The investigators, in order to obtain federal grant dollars, have to pass a hurdle in their home university or research center known as the Institutional Review Board (IRB), which studies their proposed methodology with an eye to insuring that it does no harm to the subjects. But too often the reality is that the IRB exists to protect the institution from litigation; and any protection of subjects from harm also limits the access of the subjects to benefit. In light of these realities, some social scientists are attempting a different path. They invite the "subjects" to instead become full co-collaborators. They

---

7. For example, in a little book titled *What Went Wrong?: The Clash between Islam and Modernity in the Middle East* (Oxford University Press, 2002), Bernard Lewis attributes the decline of Islam's influence over the past 1,000 years to its failure to take fullest advantage of the potential and abilities of Muslim women.

frame the studies in such a way that social scientists are embedded in the community studies, and they and the subjects are co-investigators in a common search for truth, sharing in the experimental design, the data gathering, the interpretation, and even co-authoring the papers written on the projects.

This same outlook could well be applied on a global basis. Here the number 2 enters again, raised to the power 2^{33} (which equals 8,589,934,592, a few more than the world's current population, but, barring catastrophe, we'll be there soon). Often when the conversation turns to the world's problems, it isn't long before someone sighs and says, "The plain fact is, we've just got too many people."

Yes . . . and no.

We are well versed in the "yes." Almost seven billion of us! That's a lot, however we look at it. More than Earth has ever seen before. It's more than 100 people for every square mile of land surface, and much of that surface is desert or mountain. That's a lot of mouths to feed, people to clothe and house. And, as we've discussed, per capita, many of us today are consuming resources at 50 to 100 times the rate of our forebears. Resource consumption? We're also generating mountains of waste, polluting the earth, the water, the air. And not only do we have a lot of people but we're getting a batch more, and largely in the parts of the world least able to cope.

And yet the challenge might not be the number of people per se, so much as the small fraction of us who are in a position to be problem solvers. Let's start with the some 1.7 billion people who live in abject poverty, defined as lacking basic human needs: potable water, food, shelter, healthcare. Double this figure, and you have about 3 billion people living on less than $2 per day. But all these folks are problem solvers too, aren't they? How to figure out each day how to get enough drinkable water for self and family, find food, and preserve basic decency and humanity, in a context where every person is desperate for those same things? The problems don't get any more difficult or complex than that. And the stakes? High indeed.

It's just that the problems the world's poorest people are facing are short-term, immediate, and urgent. They don't enjoy the luxury of working on the longer-term concerns. (There's another, very ugly piece to this, and that is the growing body of evidence to suggest that if you and I are short-changed in our very early development—starting in the womb and extending to those first few, vulnerable, formative years—if we don't manage to get the right nutrition and stay free of disease and, eventually, receive some education, then we never develop our fullest problem-solving potential.)

And here's some more bad news.

If we go to the other end of the spectrum, we don't find many people work-ing on long-range problems of sustainability either. How many times a day or a week are we reminded that our business leaders have a time horizon of the next quarterly earnings statement and that our political leaders are these days continually in campaign mode, strategizing and fund-raising for the next election? It often seems that leadership today is more athletic than reflective.

And what about the rest of us? We go to work each day and immediately struggle to stay afloat in a turbulent tide of e-mails. We go home 8–16 hours later not because we're through but because it's time. We're drained, spent, just from doing what it takes to get through the day. But, along the way, sometimes we've also become alienated. So instead of staying engaged in the evenings and weekends, we look for an outlet, for release, for escape. We keep on solving problems but possibly the wrong ones. Instead of focusing on our main calling in life, we're computer gaming; or doing sudoku, crosswords, and jigsaw puzzles; or reading pot-boiler novels. (You and I can substitute our personal time-wasting favorite. Advanced play? Add an overlay of guilt to these activities.)

Then there's the issue not only of what problems we're working on but what we'll accept as a solution. Keeping abreast of the blizzard of e-mails becomes a substitute for making progress on the big picture. Even when it comes to those e-mails, each day we start out with the aim of being thought-ful in each response, but at the end of each day we find ourselves tempted to just make the slightest bit of progress on the ones remaining. Our standards decline. (My parents would occasionally tell me when I was growing up that I should want a car that had been on the assembly line during the middle of the week. On Fridays the workers were looking ahead to the weekend, and on Mondays they were recovering and struggling to get their heads back in the game. Similar comments might apply to healthcare delivery and so on. There are all sorts of present-day analogies to this folk wisdom, and each of us has at one time or another, consciously or unwittingly, contributed.)

So, maybe our problem is not that we have too many people.

Rather it is that we don't have enough problem solvers, so that the prob-lem solvers are either forced or driven to work on the wrong problems and settle for partial solutions.

Fear not, this reality contains seeds of hope. Our attitude should be that every human being is a problem solver! Every solver should be focused on problems that matter. Every solver should refuse to settle for anything less than real solutions.

We can get there from here by formulating policies that foster inclusion, making elimination of poverty a priority, and investing people, time, and energy toward problems of sustainability such as economic growth, environmental protection, public safety and health, and public education. We must insist on win-win approaches at all levels from individual to local to international.

Can we get there all at once? Of course not. Each of us might consider a tithing strategy—one where we set aside 10 percent (or some other fixed level that is realistic) to make progress toward these things. And whatever you do, don't start out with a goal of maintaining such a strategy for a year. Instead, see if you can keep it up for just a week. And then decide whether you want to extend a second week. And so on. The holiday season? New Year's resolutions? A good time to rededicate ourselves along some such lines.

And finally, again in the spirit of inclusion, we shouldn't, and needn't, attempt all the necessary problem-solving alone. Rather we can seek to work collectively. We're doing this already; let's just renew our gratitude and respect for our various groups and collections and communities of colleagues, family, and friends. Let's value how the Internet is enabling more facile, more powerful, and more extensive social networking.

Before moving on, if you think 2^{33} is a big number, ask yourself how many combinations, or subsets of 2, or 3, or 20, or 30, or 200, or 300 people or so, you can form from that figure for solving problems.

Truly staggering and a reason for hope.

7.7. Critical Role of Educators and Journalists

If we're to be inclusive, society will have to develop and put into force policies that will help educators and journalists step up their game. Educators are needed because only an educated public, up to speed on the entire spectrum of complex issues, will be able to hold their elected representatives accountable and make the decisions and take the actions needed at the local level. That starts with school children in K–12. Journalists are essential because they're responsible for much of the day-to-day continuing education that the adult public receives on policy matters.

Yet, at the moment in history when they're most needed, both education and journalism are challenged. Schoolteachers are being forced to take on disciplinary roles in addition to teaching because the students' parents have failed them. Teachers are being driven to teach to standards-of-learning tests as opposed to helping their students develop critical-thinking faculties.

And journalism, the so-called "Fourth Estate," is tending to something less independent than it should be. Back in 1787, Edmund Burke was the first to voice the idea of the "Fourth Estate" to embody a political entity that was outside the three estates of "the Lords Spiritual," "the Lords Temporal," and "the Commons." He is said to have made this reference as the proceedings of the House of Commons were being opened up for the first time to journalists. Here in the United States, the term is often used to refer to the press, in distinction to the three branches of government.

Fast-forward to the year 2000. In early June of that year, the American Meteorological Society conducted its first policy forum—on the subject of hurricane preparedness and response. Meteorologists received a come-uppance at the forum when someone from or group referred to TV meteorologists and reporters as "partners" in getting out hurricane warnings and otherwise helping with emergency management. It seemed innocent enough.

But the response from one of the broadcast meteorologists was swift and cutting: "We *report* what's happening. We don't *partner* with anybody." This sensitivity to the independence of the media, and the essential importance of a free and independent press, reflected two centuries of tradition and culture. Point noted, and we went on with our session.

Yet even at that time forces had been at work for a while undermining that independence, fraying it a bit around the edges. Those trends are even more pronounced and evident today and the source of much soul-searching, not just by the press but more broadly. The once-proudly autonomous *Fourth Estate* risks slipping back to Estate 3.5.

This matters because, in all countries, the vitality and independence of the press (or their lack) shape the policy process in important ways. Can a country's leaders make deals and formulate policies in back rooms with little or no public awareness and oversight? Then that society gets one outcome (usually recognizable because graft, corruption, and malfeasance are rampant). Or are even the smallest details of the policy process open and accessible to the public? Then the people get quite another (one that most of us prefer). Incidentally, the outcome depends not simply on the press per se but also on an educated and engaged public readership. Neither is sustainable independent of the other. Their long-term fortunes necessarily go hand in hand.

So just why is the press finding it increasingly difficult to remain independent? One problem is the breakdown of the long-established business model. Let's start with newspapers. For quite some time, advertising has been the main source of their revenue. Much of that advertising came from city-center department stores and from the classified section, especially help-wanted ads. Today, fewer and fewer people are doing their shopping at major stores downtown; the action has shifted to suburban shopping malls. And most people look for job listings online. Advertisers aren't dumb; in response to these realities they've demanded lower prices. And subscription income is going south at the same time. You and I don't want to pay so much either. I get home delivery of my newspaper for less than what I paid when I first moved to Washington, D.C., over a quarter-century ago. But, arguably, I get less too. That daily newspaper is a lot slimmer now. And it relies more heavily on wire services for its content.

Newspaper staffs have been trimmed down in response. The major papers are continually cutting back staff or closing offices abroad. They're even reducing their access to major news services. Where they'd once subscribe to three—UPI, AP, and Reuters—they now make do with one or two. Increasingly, you and I are turning to the Web as a news source. It's immediately accessible and up-to-date. But it's not necessarily better, and the disconnect between the eyeballs (on the Internet) and the advertising revenues (in the hard copy) is a prescription for continuing decline.

Does the problem faced by newspapers extend to television? Yes, indeed. There was a time when we had three broadcast major networks, with centrist news staffs. Today, we have a staggering set of alternatives on cable television and online. Probably a good thing, as viewers no longer have to settle for a single centrist outlook. However, studies suggest that instead of exposing ourselves to a full range of perspectives, we tend to consult sources that play to our preexisting biases. Over time, this reinforces and hardens those views and polarizes our society. And revenues are marginal for all these sources.

One way—just a small example—of how this compromises journalist independence is exemplified with Sunday morning talk shows. To get the good ratings they require access to the highest political figures. Ask the program guests softball questions and maybe they'll allow themselves to be invited back. But to probe and make them uncomfortable, then the talk show host has just made his or her quest for the next guest more difficult. Finding the sweet spot here is not just about the news. It's also about the financial side. Again, an educated and engaged public helps. If both journalists and political figures alike know that the public insists on openness and can tell the difference between rhetoric and accomplishment, then it all becomes more sustainable.

Coverage of science and the environment has not been immune. Fewer and fewer newspapers retain environmental journalists. Increasingly they're moving to the blogosphere.

These trends are underway, just as environmental scientists are feeling growing urgency about the latest findings, and seeing a need for more concerted public action. Declines in freshwater availability and purity; growing urban air pollution; loss of habitat and biodiversity; ocean acidification—and so much more—cry out for attention. However, amid the revolutionary change underway in journalism, much of this seems to be lost in the shuffle. Announcements of these findings are either squelched or misrepresented or outright manipulated, distorted, and exploited for political gain or financial self-interest by private-sector groups and NGOs.

7.8. The Age of Renovation (Infrastructure Replacement)[8]

There's another set of skills that 21st-century humans must master, one that calls for policies to foster their development.

The discussion in section 5.1 on dams, moving our cities back from the coasts, and replacing our fossil-fuel-based energy infrastructure with something more modern and sustainable are but specific examples of a more general real-world challenge: taking preexisting infrastructure and replacing it with something more modern. Taking on and accomplishing these tasks without skipping a beat is part of the challenge of living on the real world. This Age of Renovation deserves to take its place alongside the Stone Age, the Bronze Age, the Age of Exploration, the Renaissance, the Age of Reason, or the Age of Interdependence.

Here's a quotidian example, accessible to all of us: roadways.

When my generation was young, we saw the Interstate Highway System being put in, all across America, east and west, north and south. But that system was installed some distance from the existing roads. As our parents drove, we kids could look outside the side windows of the car and see the highways being built, magnificently wide, gently curved, gently sloped, but in what seemed like a parallel universe, quite removed from the car. As we poked along, in lines of cars trapped behind aged, underpowered, overloaded trucks, snaking along hilly, curvy, two-lane roads, we could see the future, but only from afar.

Fast-forward to the 21st century. Today's driving experience, especially in and around any large city, is a shuttle through a construction zone: wall-to-wall Jersey barriers, orange-and-white barrels, traffic cones, heavy machin-

---

8. Another in the series of notional names for the age we live in. Adapted in part from a post on LivingontheRealWorld dated July 6, 2011: "NextGen and the Age of Renovation."

ery, orange signage, doubled fines. Change is occurring just as before—those new traffic lanes are being added—but not at a distance. Instead they're being introduced right on our existing roads, even as we continue our daily commute (more or less) uninterrupted. The engineering of highways is no longer just about grading and pouring concrete or asphalt and erecting overpasses; it's about doing all these things without diminishing the utility of the critical infrastructure it's upgrading. It's equal parts construction and uninterruptibility.

Highways are visible to all of us, but other infrastructure upgrades are less so. Think about the switch from analog broadcast television to digital. Or the conversion of telephone landlines to high-speed digital data transmission on fiber optics. Or the infrastructure within your house—the electrical wiring—to carry digital signals and link your new smart appliances—to the rest of your home-based IT network. Coming soon if not there already.

The point is that 100 years ago, as we were installing critical infrastructure, we weren't replacing anything preexisting. There was very little to replace. We had a virgin landscape, a blank sheet of paper, a clean slate—pick your favorite metaphor.

Today the opposite is so.

And the Age of Renovation comes at a price. It's expensive to build highway capacity on roads we're already using, versus building off to the side. It's costly to install a Metro system under preexisting office buildings or to tear up city and suburban streets to lay electrical and communications cable or water and sewage pipeline.

Here's another example: the information system that supports aviation. Called the Next-Generation Air Transportation System (NextGen), this modernization replaces radar-based aircraft flight control with GPS-based satellite technology. The current radars capture plane position only every 10 seconds or so. GPS technology enables updating aircraft position second by second. This means we can safely cram far more airplanes into the same airspace. Planes will be able to fly more fuel-efficient routes to their destination, with quicker departures from airports, more direct flight paths, and less backup at the arrival end. Fuel savings will be enormous. The passengers' travel experience will be improved.

Sounds great, right? But now, start thinking what such a changeover implies. The aircraft will need new avionics. But the early adopters won't start to see any benefit until 80 percent of the fleet has been equipped. Ground equipment will also need to be swapped out. All parties—pilots, air traffic controllers, mechanics, etc.—will have to be retrained. More flights will demand more runways. Expect litigation from counties, municipalities, and individuals over changes in the noise patterns and emissions. And what about all that military airspace? How will the new system intersect with that constraint?

The bill for the next-generation air traffic control alone totals some $30 billion to $40 billion, and that's if the renovation keeps to schedule. It's already behind. And remember, this is the Age of Renovation. Air travel can't be suspended for a decade while we put all this in place. It can't be interrupted. We will have to be able to operate under both the old protocols and the new for an extended number of years. (The bill for replacing aging roads and bridges and water and sewage systems amounts to trillions of dollars.)

Note that weather observations, products, and services for aviation play into this as well. Weather is already responsible for about half of flight delays. As we crowd the airspace, the demands on aviation weather forecasters, both government and private sector, will only increase. Our community is scrambling to keep pace with this emerging need, at a cost of more millions of dollars per year.

More such renovations are underway or waiting in the wings. Changing communications and IT continuously supplant aging technology. The military is shifting its assets from those needed to counter Cold War threats to those required to contend with globally dispersed insurgency and terrorism. Healthcare delivery needs innovation. Governments themselves could use change.

Renovation, not mere innovation. Switching out what we have for something new. Never skipping a beat. Complicated. Expensive. This is the Age of Renovation.

7.9. Transcendent Policies

In closing this chapter, three policy challenges merit special attention, because success in each respect is vital to our continuing efforts to live well on the real world.

1. *Innovation.* Innovation holds potential for achieving three goals: (1) It can buy time for the human race; (2) it can increase wealth; and (3) it can reduce the risks and costs of not knowing. Each has been discussed. Without continuing innovation, development cannot be sustained. Innovation creates resources and prevents resource issues from being a zero-sum game that pits nation against nation, institution against institution, and individual against individual. Along the way, it generates wealth, overcoming trends that would carry us the other way to aging demographics, declining ecosystem services, and increasing hazards losses. Finally, it provides our only opportunity to

avoid policy blind alleys, such as proliferation of dams, dysfunctional coastal development, and excessive addiction to fossil fuels.

Innovation policies include direct investment in science—all sciences, both natural and social. But virtually the entire policy arena affects innovation. For example, educational policy determines whether the United States, or any nation, develops the needed innovation workforce across the whole of the sciences and the national agenda. Financial policy determines whether and how the private sector invests in innovation and who benefits. Immigration policy plays into how and where the world's scientists and engineers can receive their education and training and put their skills to work.

One key question that ought to be addressed with respect to any policy on any matter is how it will foster or inhibit the innovation needed for the 21st century and beyond. And within that question is imbedded another: Will that innovation actually benefit society? Or will it merely add to the widening gap between the advance of knowledge and society's ability to benefit? One way to prevent the latter is to focus on policies that foster the coproduction of knowledge, which teams researchers with practitioners in ways that expose scientists to practical needs and expose practitioners to new ways of solving old problems. And actually that characterization trivializes what really occurs in such collaborations. As collaborations between practitioners and researchers develop, the distinction between the two dims. They each become something more.

2. *Learning from experience.* Chapter 7 provides an illustration of how learning from experience works in the instance of the National Transportation Safety Board and suggests that an analogous process be established for learning from experience with natural disasters.

But given the urgency and demands of our 21st-century predicament, learning from a limited set of experiences is not enough. We need to adopt policies that will greatly expand the number of experiences from which we can learn. This is the main focus of Chapter 9, but a few words are in order here. The key policy challenge is to devolve responsibility for innovation and for learning from experience from the level of the nation-state down to the local level. This has the effect of increasing the number of policy experiments that can be run per unit time. It also implies that policies have to foster early detection of success or failure. Remember, policies have emergent consequences that are difficult to appreciate fully prior to actual experience.

3. *Funding for policy research itself.* This requirement implies, inter alia, a need to build capacity for policy analysis at all levels of society. This has proven problematic in the past, as political leaders have seen policy analysis

as a weapon that can be used against them. But as later chapters of the book will make clear, the need for improved, more rapid, more adaptive, and more effective policy analysis and formulation is as great for the United States in the early 21st century as the country's need for policy for basic research was back in the Vannevar Bush days immediately following World War II. At the national level, we have three success stories and one setback. The success stories are (1) the Congressional Research Service (CRS), which provides nonpartisan analysis to members of Congress on any and all issues in response to direct request, (2) the Congressional Budget Office (CBO), which analyzes the economic impacts of federal legislation ranging from tax policies to healthcare policy and everything in between, and (3) the Government Accountability Office (GAO). The setback was the Office of Technology Assessment (OTA), which was disestablished in 1995. In our government, with its separation of powers and checks and balances, Congress needs every bit of this analytical capacity to ensure a level playing field with the executive branch in the formulation and execution of policy. The outlook, which is for increasing demands on the policy arena across a broad spectrum of national and international concerns, suggests that Congress could do with more.

Real-World Living Initiative #3. The United States should build the capacity of CRS, CBO, and GAO, and at the same time reestablish OTA. There should be some high-level consideration of ways and means to fund academic research in policy analysis in a more structured, formal, robust way, and in the process educate and train a generation of analysts who could employ those skills at state and local levels where they are badly needed, and who could provide the needed diversity and disciplined thought with respect to policy formulation.

7.10. Recap

Policies are potentially vital tools for translating scientific knowledge and understanding into societal benefit, but they can and should be improved. This chapter is only one of many possible enumerations, but it's more than an inventory. It's a to-do list. No matter who you are, where you live, or what your age and experience and/or educational background and workplace are, you can start improving on one or more of these aspects of policies and their application, while staying in place. And because most policies' impacts are emergent and viral, you can in this way greatly leverage your lifetime contribution to making this a better world.

The conversation here may have seemed high level and general to some. Partly for that reason, the next chapter hones in on just one policy aspect, to illustrate the potential.

8

HAZARDS POLICY INNOVATION

Civilization exists by geologic consent, subject to change without notice
—*Will Durant*

Chapter 7 provided a quick, high-level overview of areas where policy is in need of analysis, reformulation, and further development. In this chapter we zoom in on hazards policy and offer some notional ideas for improvement in this limited context. Some of the ideas are original; others are borrowed.

Recall from Chapter 6 that policies derive much of their power and effectiveness from their emergent properties, impacts that may not be self-evident from first reading but appear as the policies are put into practice. Note as you read that none of the suggestions here appears to tackle the problem of hazards vulnerability head on; rather they come at it obliquely. The suggestions here represent very small, inexpensive, quick adjustments we could make to the ways we do business. The assertion is that, taken separately or together, they'd lead to substantial reductions in hazards losses, not so much directly but rather through their emergent consequences over time.

Two reasons underlie the emphasis here on natural threats. First, the hazards piece is the one with which I'm most familiar, especially hazards policy in the United States. But second, much public policy suffers from the fatal flaw that it fails to recognize we're trying to achieve our policy goals on a planet that does its business largely through extreme events.[1] Thus we'll explore this element in a bit more depth, not because it's any more important than resources policy or environmental policy but because it affords us a look at some overlooked opportunities for substantial improvement.

As said, because we live on a planet that does its business through extreme events, natural hazards appear in the policy framework of every nation. As it happens, when we look worldwide, the United States is arguably the country with the most to lose. With dozens of volcanoes, a lot of territory on seismic fault zones, as many hurricanes as China, as many winter storms as Russia or Canada, and a virtual lock on the world's tornadoes, the United States should see it essential to get hazards policies right. And that's before we factor in U.S. global interests. In unique ways, the United States is interested in the impacts of hazards worldwide. The tragic loss of life and human suffering, property destruction, and economic disruption from natural hazards aggravate preexisting social strains in every country. They lead to social unrest, and in the worst scenarios to mass migration of displaced populations, in some cases across national borders. These events lead to geopolitical instability that is harmful in and of itself but also threatens

---

1. My colleagues at the AMS Policy Program, being rather more disciplined than I am, occasionally grow restive at my loose use of the word extreme in this and other contexts. My apologies to them and to you. What I mean is probably best captured in the spirit and far greater precision and eloquence of Nassim Nicholas Taleb's book *Black Swan: The Impact of the Highly Improbable* (Random House, 2007). Taleb notes that many investors attempt to make money in small amounts on a daily basis in the financial markets using approaches that may appear to be working 9,999 days out of 10,000 but lose more on that next single day than they've accumulated in all prior years. I'm simply acknowledging that, in big, chaotic systems such as Earth's atmosphere and oceans, the averages are very often the result of relatively large, competing swings in both directions from the mean. Here, as on the financial markets, the largest swings contribute disproportionately to the averages. For example, at many if not most locations worldwide, the total rainfall for a given month, or an entire season, may result from just a few days of rain during the period, the passage of a single hurricane, or one winter storm. The direct climate change threat to island nations from sea level rise may prove to be small compared with the threat posed by concomitant changes in hurricane patterns of track or intensity, which may bring storm surges that are brief in duration but that may inundate islands previously thought out of harm's way. For one day out of 10,000, the purpose of a home's walls may be to hold the roof down despite high winds versus support the roof against the pull of gravity. And so on.

strategic interests of other countries, including China, India, Brazil, Russia, and other countries in Europe and the Americas.

Recall from Chapter 4's persistence forecast that we are tending globally to a zero-margin society. Almost by definition such societies are more vulnerable to extremes of nature: disruption of national and global food supplies and more. Despite the high stakes, the United States has been slow to think strategically about the threats posed by natural hazards. Instead the policy approach has been largely confined to (1) attempts to manage emergency response of increasing scope, as disasters and their impacts spread, and (2) the provision of forms of insurance.

As coastal populations have grown, emergency response, especially evacuations, has proven particularly problematic. The historic experience is that only something like 50 percent of those in populations ordered to evacuate do so, while as many as 50 percent of those who do evacuate are people who've specifically been asked or encouraged to shelter in place. (Statistics for Hurricane Katrina were substantially better, but we have yet to see whether this was an isolated instance or the start of a trend.) Recent events have seen as many as six million people taking to the road in affected regions. In light of the fact that many are reluctant to leave because of concerns for the safety of their property while they're gone, and many of those who do leave return to find damaged homes, disrupted infrastructure, and lost jobs, it would seem the country should place greater, more strategic interest in reducing the scale of evacuations needed rather than relying narrowly on managing evacuations of increasing scale and complexity.

Efforts to provide government insurance against natural hazards—as represented, for example, through the National Flood Insurance Program, through various forms of government reinsurance such as so-called catastrophe bonds, indirectly through political tinkering with insurance rates at the state level (forcing insurers to price protection against earthquakes and storms at less than that warranted by the actuarial risk), or through after-the-fact bailouts—have also proven misguided.[2]

Here are six suggestions for improving hazards policy. If adopted they'd likely substantially reduce losses over time.[3]

---

2. Crop insurance might also fall in this category.

3. Most of the ideas that follow here appeared first in posts on LivingontheReal-World between September 2, 2010, and September 9, 2010. I've returned to variants on these ideas many times and in many forms since then on that blog.

8.1. First, Do No Harm

The origin of the precise phrase hasn't been traced, but the sentiment of *non-maleficence* (what a wonderful word!) is embodied in the Hippocratic Oath and is the starting point for medical ethics. It is often invoked as a broader rule, for governing our actions throughout every aspect of our lives. We can all embrace the idea.

Or maybe not. It turns out that we sometimes have a bit of a blind spot when applying this principle to natural hazards. One organization, the Association of State Floodplain Managers (ASFPM), has been quite articulate on this point for several years. They term this principle "no adverse impact" (NAI), and they propound the idea with a lot of solid material on their website. Here's a simple example that will make the point. Let's say that you live in a town on one bank of a river and that there's also a town on the opposite bank of the river. Both towns experience repetitive loss to floods. Let's say that both towns decide to work with the Corps of Engineers (or whomever) to build levees for an added measure of protection. If both towns build levees to the same height, you still share the residual risk from floods exceeding that height. But then it occurs to you: If your town builds a levee a foot higher than the town across the way builds its levee, your flooding problem will be solved.

Is that fair? Should that be a decision you're allowed to reach on your own? Of course not.

Now think on it just a bit more. When both towns do build levees to the same height, the two towns together have basically exported the flood hazard to all the towns downstream. Have the inhabitants and the business leaders in those downstream towns been consulted? If not, that doesn't sound quite fair either, does it?

That's because it's not.

In other areas of the law, you and I are to operate under the legal admonition "*sic utere tuo ut alienum non laedas.*" (In plain English, "so use your property that you do not injure another's property.") This means as landowners we can't be or create a nuisance to our neighbors, impair their rights, trespass, and so on. But the courts and other players in the legal system, thanks in large part to ASFPM, are now working out how this applies to floodplain management. (Note that as their name implies, ASFPM doesn't advocate abandoning the idea of protection against floods; they simply posit that there are complements and alternatives to levees as a policy option and that the levee option itself requires careful thought.)

But why confine this principle to the flood hazard alone? Shouldn't, or couldn't, it apply to hazards more broadly? For example, should I be allowed to leave debris lying around in my yard that could become windborne in a tornado or hurricane and damage my neighbor's house? Or fail to fireproof the shingles in my roof in a wildfire-prone area and thereby endanger my neighbors' homes? A relatively new organization, the Natural Hazard Mitigation Association (NHMA), is currently formulating such arguments. The NHMA is interested in part because of the inherent logic of the case; in part, because one of its officers, President Ed Thomas, was himself the author or coauthor of much of the ASFPM material.

Some readers may have a sense of déjà vu. Environmental Impact Statements (EISs) embody a similar principle. The National Environmental Policy Act of 1969, prompted in part by the Santa Barbara oil spill, required federal agencies to assess and report on the environmental consequences of their contemplated decisions and actions. Hotly debated at the time, EISs are now a widely accepted way of doing business.

The same can happen in the context of hazards policy over time.

Reflect on this. Environmental impact statements are of value in and of themselves, but perhaps their greatest contribution was in building public awareness, at all levels, of (1) the consequences of our actions on habitat, biodiversity, air, water resources, and soil, and (2) the idea that in the process of growing more interdependent, we increasingly have to balance our individual freedoms against our responsibilities to others. In the same way, if all of us can join ASFPM and NHMA in backing the no-adverse-impact idea as one cornerstone of our hazards policy, we'll become far more effective at living on a planet that does its business through extreme events.

8.2. Learn from Experience

Mistakes are a fact of life. It is the response to error that counts. —*Dag Hammarskjold*

Figure 8.1 is a bar chart showing world losses (insured and uninsured) to natural hazards over the past century or so. The statistics are based on data collected and analyzed annually by Munich Reinsurance. The losses are noisy—highly variable from year to year—but the long-term trend is evident, even without the dashed lines.

The trends shown reflect the dollar losses *after* any effects of inflation have been removed; they're growing at about 6 percent a year on average, comparable to or greater than the growth in world gross domestic product (GDP).

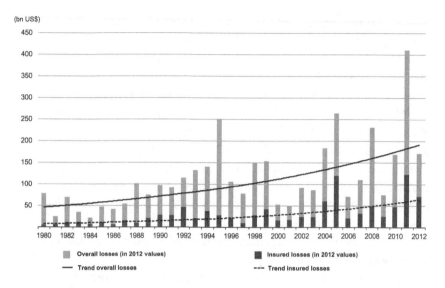

FIGURE 8.1. World losses from natural hazards. © 2013 Münchener Rückversicherungs-Gesellschaft, Geo Risks Research, NatCatSERVICE.

Ask experts for an explanation, and most will tell you that populations are generally rising; people are increasingly settling in more hazardous areas, like the world's coasts; and they're gradually getting richer, so that the property exposure per capita is increasing. The implication is that rising losses can't be avoided.

But is that necessarily inevitable? Let's compare this trend with a look at commercial airline accidents, over a comparable period of years.

Commercial flights per year have increased perhaps tenfold since 1960. However, there's no corresponding increase in fatal accidents. The number of accidents varies quite a bit from year to year (Fig. 8.2), but if anything the overall trend is slightly declining. Why aren't commercial aircraft accidents dramatically increasing? If the accident rate today were the same as 1960, we'd be seeing 10 commercial airline disasters each week, more than one a day.

Let's go back to the natural hazards trend. In 2001, a famous trio of social scientists—Gilbert White, Ian Burton, and Bob Kates[4]—offered a different perspective and explanation for what is going on there. In an article titled

---

4. White, G. F., R. W. Kates, and I. Burton, 2001: Knowing better and losing even more: The use of knowledge in hazards management. *Global Environ. Change*, **3B**, 81–92.

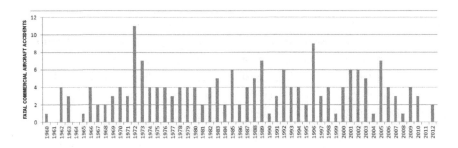

FIGURE 8.2. Fatal commercial aircraft accidents by year, courtesy planecrashinfo.com.

"Knowing better and losing even more: The use of knowledge in hazards management," they concluded that (1) knowledge is lacking, (2) knowledge is available but unused, (3) knowledge is available but used ineffectively, (4) there is a time lag between the application of knowledge and the results, or (5) the best efforts to apply knowledge are overwhelmed by the rapid increase in vulnerability. (Their article goes into a lot more detail and is thus worth a serious read.)

That matches with the image we see on TV in the aftermath of a natural disaster: Networks show a man, a woman, or a couple at the scene of what's left of their former home. They're picking through the wreckage and are (quite appropriately and sadly) distraught. They're grieving. They're saying things like "this was my (our) grandparent's home, and we're going to rebuild as before."

By contrast, in the aftermath of an aircraft accident, the images show the wreckage of a crashed plane (and this is a relatively rare occurrence, by the way). Someone, perhaps wearing a National Transportation Safety Board (NTSB) jacket, is saying, "We're recovering the flight recorder and collecting all the evidence. We're going to understand the reasons for this accident. We can't let this ever happen again."

Wow. These video clips certainly seem to be worlds apart. When it comes to natural hazards, there's a drive to rebuild as before. This shows courage; it shows a triumph of the human spirit (measured one way); and it shows "the importance of place," as we all know in our gut and as social scientists are quick to remind us. All of this should be respected. However, viewed through another lens, it can also look like a failure to learn from experience. (There's another huge failure here; when our generation does in fact rebuild as before, it's not entirely an act of courage. We're pretty much condemning future generations to that very same loss some year in the future.) Now let's look at commercial aviation. If people heard someone saying "The wing fell off this airplane, but we're going to rebuild this plane as before," they would walk off shaking their heads and vow never to fly again.

The vignettes show two different sides of our individual (hopelessly split?) personalities. Why do nations and states—and, for that matter each of us— respond so differently to these two scenarios?

Ask people about this, and you'll get a range of answers. Some say that in the case of natural hazards, it's the sense of place that's the driver. The attachment to our roots is so strong as to trump everything else. Some blame programs such as the National Flood Insurance Program (NFIP). They say flood insurance creates a moral hazard. They cite cases of repetitive loss like the Wilkinson, Mississippi, home that has been flooded 34 times since 1978, a home currently worth about $70,000 but racking up insurance payments of $660,000 over the period. Some say the natural hazards case is more complex, more politically charged, than the aircraft accident. Obviously, many different motivations contribute.

But it might be in part that the NTSB is playing a critical role. That's why, in 2006, writing in the *Bulletin of the American Meteorological Society*, my colleague Gina Eosco and I suggested that the country needs an analog to the NTSB—a National Disaster Reduction Board (NDRB)—with a mandate to investigate the full range of disasters: natural hazards, including disease outbreaks; industrial accidents; and willful acts of terror.[5] The existing NTSB has a number of features that make it an interesting model. It is small, only about 400 people. It is independent (contrast the recent dust-up following the BP Oil Spill of 2010 about the failure of the Minerals Management Service to keep the oil industry it was nominally supervising at sufficient arm's length). It only investigates causes of accidents and issues findings, leaving it up to the FAA, the airlines, the airframe manufacturers, and others to implement or ignore its suggestions. It brings on representatives from these stakeholder groups (the airframe manufacturer and/or the airline[s] involved) into its investigative teams. (At first blush, this might seem inconsistent with the arm's-length idea; however, the employees of the stakeholders sign very specific legal agreements governing their conduct while on such NTSB teams, limiting their contact with their employers, for example.) It investigates all possible causes or contributors to the incident. For example, when the wing falls off a plane, the NTSB doesn't confine its attention to pilot error but

---

5. Eosco, G. M., and W. H. Hooke, 2006: Coping with hurricanes. *Bull. Amer. Meteor. Soc.*, **87**, 751–753.

looks instead at wing and plane design, maintenance and operation, weather conditions, and so on.

Perhaps equally important is that the NTSB is a standing federal agency. For a number of years, the National Science Foundation used to fund the National Academies of Science–National Research Council (NAS–NRC) to do postdisaster studies on an as-needed basis. Experience showed that the academic committees that did the investigations produced excellent products but sometimes only after unacceptably lengthy delays. In more recent years—following 9/11, Hurricane Katrina, and other such events—Congress has established special commissions to investigate. Sometimes, because these groups are ad hoc, they are forced to spend considerable time agreeing on an investigative process. (By contrast, the NTSB operates under tried-and-true protocols.) In the case of Hurricane Katrina, the White House, the House of Representatives, and the Senate mounted separate studies, each numbering hundreds of pages. However, each of the three reports focused only on the shortcomings of the emergency response during the few days around hurricane landfall. They ignored a broader set of political and engineering decisions at federal, state, and local levels regarding land use, construction of levees, and resettlements of populations and businesses that were made over periods of many years.

(A modest effort along these lines is being explored internationally, as part of an International Council for Science [ICSU] program on Integrated Research on Disaster Risk [IRDR]. The IRDR includes a component called Forensic Investigations of disasters [FORIN].)

8.3. Keep Score: Estimate Losses Year to Year

In the early 1900s, my grandfather faced a challenge at work. Though only a teenager, he was foreman in a foundry making cast-iron bathtubs in Chattanooga, Tennessee. His company was struggling. A large number of the bathtubs they produced were defective, so badly flawed they had to scrap them. They were losing money. What to do?

My grandfather was a baseball fan. He solved the problem the way a baseball fan would. He got a big blackboard. He hung it on the foundry wall. He wrote every workman's name on it. Next to each name he started keeping a tally: How many passable bathtubs had that worker produced that week? And what was his batting average? Of all the workers, who was the best that week? The Top Tubber? The MVP?

The workers reconnected with their competitive side. Almost overnight, the foundry's output shot up. Defects went down. No one had to be threatened with loss of a job. No one had to be offered any more pay. Morale improved. All that was needed, as it turned out, was a scorecard.

Maybe we can scale this up. If we want to reduce disaster losses, why shouldn't we start by keeping score? We certainly do this in other areas. Take the economy. Examples are manifold: GDP growth, month by month, or year by year; unemployment statistics; balance of trade; housing sales; inflation; consumer confidence. These and other figures make the news on a regular basis. They drive financial markets. They trigger public discussion and debate. And as they do, they help us improve national economic performance. Increasingly, such figures drive the environmental discussion as well: the number of high-ozone days per year in cities; fish stock statistics; global temperatures; atmospheric carbon dioxide concentrations and emissions. And it doesn't stop there. Metrics, measures of success, performance and results, and indicators permeate the whole of the public policy discussion. OSHA tracks accidents in the workplace, seeking to reduce them. Every year, in all public schools, we'll test students and compare their reading and math scores with kids from other schools in the United States and abroad. It is a preoccupation both nationally and internationally. The nomenclature changes from field to field, or from administration to administration, but the idea is the same. We hope to get better at things we measure.

No idea is completely new. This notion, or something like it, has already come up in the hazards arena. Take the StormReady Program run by the National Weather Service (NWS), where counties, communities, Indian nations, universities, and other groups or organizations can all qualify by meeting certain criteria. For the most part, these criteria focus on preparedness. How alert is the community in question to NWS warnings? How ready are they to respond? The Institute for Business and Home Safety, a group of insurers, has its Fortified program, which recognizes homes that have been constructed to a higher, safer standard than the normal building codes. Munich Reinsurance annually tallies disaster loss figures.

All well and good. But it reminds me of another family vignette: When my dad was in his late seventies, he had a blood workup at the clinic. Looking at the lab results, the doctor said, "Mr. Hooke, you have the blood of a 26-year-old!" "That proves," my father replied, "that you're measuring the wrong things."

It's also important to measure not just what is easy to measure but what matters most. Suppose, on that 1900s blackboard, my grandfather had tallied attendance? Or who arrived earliest? Or stayed the latest? Performance in those areas would have improved. And it might well have helped the foundry but only indirectly. Something similar happens in baseball. Statistics tempt athletes to focus on their individual batting averages or runs batted in. But over the past half-century, managers have started refining those measures. What happens to a player's batting aver-

It's also important to measure not just what is easy to measure but what matters most.

age when men are on base? How about RBIs when the game is still in doubt? And what really counts is not the number of runs but the "W." That's the bottom line for the team.

When we apply this test to hazards and the above examples we see that some work still remains. It's important for communities to be ready to re-act to oncoming storms. But maybe we should be engineering our homes and towns so there's less need for evacuation or taking special shelter. The Fortified program addresses this to some extent, but it doesn't address the community-wide performance of critical infrastructure and the extent of business disruption. How should we capture these dimensions in our score-card? Consider the drought and heat wave of 1980. Water levels were so low in the Mississippi River basin that barges couldn't ply their trade. The result was a spike in the cost of getting coal to utilities all through the Midwest. Farmers and ranchers in the southern tier of states lost something like $20 billion of livestock and poultry. But as one economist told me at the time, this didn't show up in national GDP figures, because farmers in the northern tier of states received higher prices for their cattle and chickens that year. So, in terms of the national accounts, there was little or no net loss to the economy. But there was a $20 billion transfer payment from the southern states to the northern states. And GDP was an inadequate measure of the turmoil and disruption for many Americans that year. (Something similar happens in flooding or windstorm damages. The property loss may be compensated in part by a construction boom.)

The last serious look at these questions—what to measure? how? why?—dates back more than a decade. In the late 1990s, under the auspices of the NAS–NRC, Robert E. Litan of the Brookings Institution led the Committee on Assessing the Costs of Natural Disasters. They wrote a report titled "The Impacts of Natural Disasters: A Framework for Loss Estimation." The find-ings and recommendations included a look at the diverse kinds of direct loss that are incurred, a call for greater attention to indirect losses such as unemployment and business disruptions, encouragement for more uniform, standardized measurement of losses across different hazards, and much more.

The Committee recommended that the Bureau of Economic Analysis of the U.S. Department of Commerce, working with other federal agencies such as FEMA, might take the lead in establishing the needed database. They noted: "[R]esearchers and experts in disaster loss estimation could benefit from a standardized data base that would enable them to improve estimates of both the direct and indirect losses of disasters. *These improvements in turn would assist policymakers in their efforts to devise policies to reduce the losses caused by future disasters* [emphasis added]."

8.4. Foster Public-Private-Sector Collaboration

To bring down the cost of future disasters, instead of relying on public-sector leadership alone, we can team the public and private sectors as full and equal partners in the effort.

Isn't this going on already? Yes and no. Let's go over a little background. Before we dive in, here's something to chew on that ought to motivate us. These days, historians, social scientists, and policy analysts sometimes ask why certain societies and cultures fared better than others over the past 100 (or 1,000) years. One explanation keeps cropping up. The societies that do the best tend to be those that draw most successfully on their full population, and not half of it—those who grant equal opportunity to women throughout their culture, but especially in the workplace. Now, when it comes to hazards, reflect that in the United States we try to cope with the threat of natural hazards using not *half* the workforce *but only 10 percent*: the public sector, those employed by federal, state, and local governments (and obviously, only a minority of these; most government workers play roles quite distant from risk management).

Of course the private sector is already involved. They're threaded throughout the process in a number of ways. Here's a partial list.

1. *Victim.* Analysts tell us that something like half of the small businesses who are able to stay open, who have to close their doors during a disaster, never reopen. In Hurricane Katrina, just to choose one instance, half of the local businesses had to shut down. No wonder the regional recovery has been so slow.

2. *Vector.* Just as mosquitoes are a vector for the spread of malaria, globalized private-sector firms spread the effects of once-local disasters throughout entire regions and worldwide, through their effect on distant suppliers and customers.

3. *Emergency responder.* It wasn't always this way, but nowadays private-sector contractors work in emergency operations centers in cities, counties, and states around the country, as well as in federal emergency management agencies. Radio and television broadcasters, all in the private sector, are a primary means for getting emergency information to the public. Hospital emergency rooms are often in private hands.
4. *Critical infrastructure provider.* State and local governments may control most police and fire departments, most roads, schools, and sewage infrastructure. But the private sector controls much of the electrical power system, communications, gas pipelines, hospitals, the financial system, and so much more.
5. *Recovery partner.* The private sector does much of the rebuilding and restoration of community function.

It's a pretty extensive list! So how can we possibly say the private sector is shut out? In what respect?

How about as *strategic partner*? Of course, there's a reason for this. In the United States, government and private enterprise have for the most part profited from a fundamentally arm's-length approach, in which the private sector is free to make a profit and the government uses regulation as a tool to ensure that when all parties act in self-interest, they also contribute to a common good. When they do work together, it is usually (excuse the legalese) as principal and agent (governments determine what's to be done; the private sector provides the desired product or service as a means to that end).

But the United States could quite possibly enjoy much greater business continuity, and far less community disruption, if government put into place policies that would encourage public–private collaboration.

But the United States could quite possibly enjoy much greater business continuity, and far less community disruption, if government put into place policies that would encourage public–private collaboration (and expanded the notion of the private sector to include nongovernmental organizations [NGOs] and faith-based organizations [FBOs] at the local and national levels).

This was tried, briefly, during the 1990s under the leadership of FEMA Director James Lee Witt, who established Project Impact. Under this program, the government gave selected communities small planning grants on a competitive basis, to stimulate strategic cooperation between local governments and businesses, and to focus that strategic-level cooperation on pre-event measures to reduce or mitigate community risks, as opposed to merely responding to and recovering from disasters. The program proved so popular that when the funding ran out, many communities asked to be included under the umbrella at no cost, providing the resources themselves. The program showed promise, but it was viewed as politically motivated and following the 2000 elections was discontinued.

Whether under this or some other label, the concept deserves another chance.[6]

8.5. Revitalize a Venerable Institution

Interestingly, the U.S. Department of Commerce might be able to play an important role here, building on its own ties to the corporate world but also on an extensive suite of in-house capabilities to support business continuity. These include but are not limited to the following:

- NOAA (National Oceanic and Atmospheric Administration), which provides many hazard outlooks and warnings, for weather and also for tsunamis and volcanic ash plumes;
- NIST (National Institute of Standards and Technology), which supports extensive wind, fire, and seismic engineering research and development;
- EDA (Economic Development Administration), which helps rebuild the local economy, just as FEMA tries to maintain family life, by housing disaster survivors in a trailer;
- Census, which keeps an inventory of vulnerable populations; and
- ITA (International Trade Administration), which could help U.S. consulting and engineering firms find international markets for their expertise in disaster reduction.

A helpful partner here would be the Business Civic Leadership Center (BCLC), embedded within the U.S. Chamber of Commerce. A number of

---

6. A few years ago, I chaired a NAS–NRC committee that looked into this issue in considerably more detail (chaired is a fancy word for struggling to keep pace with a dynamic committee and a go-getter NRC staffer who provided adult supervision). Our report, "Building Community Disaster Resilience through Private-Public Collaboration," was published by the NAS Press in 2011 and is available online.

major corporations have banded together to provide a wide range of help to businesses and communities worldwide struggling to prepare for or recover from disasters, working with Stephen Jordan, their very effective executive director, and his small but immensely talented staff. Or look at Business Executives for National Security (BENS), another NGO looking at a closely linked set of risks. These and similar organizations hint at the depth of private-sector interest and capabilities with respect to these matters.

My colleagues in the hazards policy arena could add much more texture. There's opportunity for progress in revisiting programs such as the NFIP, the National Earthquake Hazard Reduction Program, crop insurance, and many more.

8.6. An Aside: Recognize That "Disaster Recovery" Is an Oxymoron[7]

A few years ago I was very kindly invited to give a keynote address to a workshop of top hazards experts convened for the specific purpose of developing a theory of disaster recovery or perhaps more modestly to lay out a notional program of research that might over time lead to the emergence of such a theory. They were kind to let me in, but they may have been badly served by their generosity. For one thing, I've been in the hazards community a long time, but I'm only a grafted branch. I don't really have a research pedigree in the field. Anyone in the room, chosen at random, would have been better qualified to give the keynote. (In fact, I'm rather sure, and may even have been told, that I was selected for precisely that reason. The meeting organizers knew that no one sitting in the room would be jealous; they would all recognize that the keynoter had been chosen for reasons other than competency.) For another, I've usually been happier summing up at the end rather than starting off. It's easier and more effective to first seek to understand and then be understood than the other way around. And finally, and most fundamentally, I'm pretty sure that disaster recovery as I would define it (not as they would define it) was an oxymoron—and I told them so.

One obstacle to clear thinking about disasters is the prevalent use of soothing nomenclature that encourages a false sense of security among policymakers and the public. One such phrase is "disaster recovery."

There is no such thing. Disasters are in reality events that exceed the critical limit of community resiliency, beyond which true recovery is impossible. Here's an analogy to help see this. Hang a weight on a spring, or

---

7. As discussed in Chapter 7. the material here was largely excerpted from posts on LivingontheRealWorld, dated March 18, 2011, and January 3, 2012.

open a paper clip and it will stretch to accommodate the need. Take off that weight or remove the sheaf of papers and the spring or paper clip will return to its original shape. But put on too much weight, and the spring will permanently deform. It has experienced a disaster. It will never recover its original shape or utility.

But when it comes to hazards this is not what we're told.

It starts with the notion of *resilience*. Some years ago, Fran Norris and her colleagues at Dartmouth Medical School wrote a paper that has become something of a classic in hazards literature.[8] They introduced the notion of *community resilience*.

Here's the abstract for their paper. As you read it, reflect on what you've read and remember about September 11, 2001, and the World Trade Center; Hurricane Katrina and New Orleans; and the Sendai earthquake and tsunami (which the U.S. Geological Survey has named the Tohoku earthquake). Think about how this description of community resilience bears on those events.

> Communities have the potential to function effectively and adapt success-fully in the aftermath of disasters. Drawing upon literatures in several disciplines, we present a theory of resilience that encompasses contemporary understandings of stress, adaptation, wellness, and resource dynamics. Community resilience is a process linking a network of adaptive capacities (resources with dynamic attributes) to adaptation after a disturbance or adversity. Community adaptation is manifest in population wellness, defined as high and non-disparate levels of mental and behavioral health, functioning, and quality of life. Community resilience emerges from four primary sets of adaptive capacities—Economic Development, Social Capital, Information and Communication, and Community Competence—that together provide a strategy for disaster readiness. To build collective resilience, communities must reduce risk and resource inequities, engage local people in mitigation, create organizational linkages, boost and protect social supports, and plan for not having a plan, which requires flexibility, decision-making skills, and trusted sources of information that function in the face of unknowns.

Here's some more material on the same general idea, taken from a website called *learningforsustainability.net*: "Resilient communities are capable of

---

8. Norris, F. H., S. P. Stevens, B. Pfefferbaum, K. F. Wyche, and R. L. Pfefferbaum, 2008: Community resilience as a metaphor, theory, set of capacities, and strategy for disaster readiness. *Amer. J. Community Psychol.*, **41**, 127–150.

bouncing back from adverse situations. They can do this by actively influencing and preparing for economic, social and environmental change. When times are bad they can call upon the myriad of resources [sic] that make them a healthy community. A high level of social capital means that they have access to good information and communication networks in times of difficulty, and can call upon a wide range of resources."

Taking the texts pretty much at face value, as opposed to a more professional evaluation, do you recognize "resilience" in the World Trade Center, Hurricane Katrina and New Orleans, the Tohoku earthquake and tsunami, and Hurricane Sandy in this framing?

Maybe, yes, and no. No, if you zoom in and look at the pre-2001 Twin Towers, or the individual small Japanese towns and neighborhoods entirely obliterated by the tsunami, or the Fukushima nuclear plant in isolation. They're through. Finished. Other communities and other electrical generating plants may come in and take their place. They may take the same names. But they'll really be entirely different. To call that recovery won't really honor or fully respect those who lost their lives in 9/11 or the Japanese flood and its aftermath.

To see the resilience of which these texts speak, you have to zoom out, step back quite a ways. The smallest community you might consider? That might have to be the nation of Japan in its entirety. And even at that national scale the picture is mixed. It'll take decades to sort out the longer-term implications for Japan.

This is happening more frequently these days. The most recent Eyjafjallajökull volcanic eruption, unlike its predecessors, disrupted much of the commerce of Europe and Africa. In prior centuries, news of the eruption would have made its way around the world at the speed of sailing ships, and the impacts would have been confined to Iceland proper. Hurricane Katrina caused gasoline prices to spike throughout the United States, not just the Louisiana region. And international grain markets were unsettled for some time as well, until it was clear that the port of New Orleans was fully functional. The "recovery" of New Orleans? That's a 20-year work-in-progress. It'll be many years before the coastal areas of New York and New Jersey return to a new normal.

And go back just a little further, to September 11, 2001. In the decade since, would you say that the United States functioned as a resilient community, according to the above criteria? Have we really bounced back? Or have we instead struggled mightily with "build(ing) collective resilience, communities . . . reduc(ing) risk and resource inequities, engag(ing) local people in mitigation, creat(ing) organizational linkages, boost(ing) and protect(ing) social supports, and plan(ning) for not having a plan, which requires flexibility, decision-making skills, and trusted sources of information that function in the face of unknowns"?

Sometimes it seems that 9/11 either made us brittle or revealed a pre-existing brittleness we hadn't yet noticed—and that we're still, as a nation, undergoing a painful rehab.

All this matters because such events seem to be on the rise, in terms of impact and frequency. They're occurring on nature's schedule, not ours. They're not waiting until we've recovered from some previous horror, but rather they're piling one on top of another. The hazards community used to refer to these as "cascading disasters." Somehow the term seems a little tame today.

And you'll note that what we refer to as "community resilience" really refers more to a community that has been struck a glancing blow. Think of resilience as healing. A soldier loses a limb in combat. He's resilient and recovers. A cancer patient loses one or more organs. She's resilient and recovers.

But the severed limb never does. The excised organs never do.

In the same way, community recovery rarely means recovery of the original people, or the original homes, or the original businesses. Suppose a town experiences a flood, and some people lose their lives and other people are injured. Some homes are swept away, and some businesses permanently shutter their doors because they've lost their customer base.

Now watch what happens over time. After a little while, some new homes are built on the sites of the old ones. Some of the original folks who survived maybe move back, but also a few new people move in from somewhere else. Soon those shops on Main Street are open again, but many house different businesses. Wait a little longer, and still more new homes appear, and more people, and more business. Eventually do a census, and discover that the town is back to its "original condition." Do this for a number of communities, and you'll find they take different amounts of time to return to a similar state. So you might be tempted to say that some are more resilient than others.

So often it seems that the community resilience we speak of is really the resilience of some larger community or region that wasn't so affected by the flood (or other disaster). But that's not how we speak of it.

But this distinction matters; it is more than just splitting hairs. The "new normal" is likely to refer to a far different future. Call to mind the eruption of Mt. St. Helens. In the weeks and months immediately following the eruption, the terrain was a sterile landscape buried under feet of ash. Today it's green again, but not with old-growth forest. It's a different blend of plants and critters, and it'll evolve differently from this point on. Think back to Hurricane Katrina and New Orleans. Much of the black community left and has yet to return. The influx of Latinos has been significant. The demography and destiny of the city have been forever changed.

The difficulties are tragically underestimated.

People who haven't experienced disasters personally are often unworried about natural hazards. They feel lucky. They don't think the hazard will hit them. They trust that their local officials have been conservative about such risks when regulating land use and building codes. They believe they'll get warnings in time to seek and find shelter. They think their city or country or state has adequate plans in place for contingencies. They figure that federal agencies such as FEMA, SBA, and others will make them whole.

Such confidence in national, state, and local institutions is naïve and misplaced. As a country and a world, we'd likely be better off if we stopped sugar-coating the consequences of failing to cope with natural hazards—if we used the idea of resilience to refer to the capacity to bend in the face of extremes without breaking versus "recover" from breaking.

8.7. Invest in Seed Efforts to Anticipate, Forestall Global-Scale Catastrophe

This holds further ramifications for us. If resilience is not an intrinsic community property but instead exists only because there's a larger community outside a finite disaster zone that was largely unperturbed, then as the scale of disasters continues to increase and time moves on, increasingly we will encounter or experience disasters of such large scale that there are no unaffected populations or regions to be resilient. Recovery in such regimes will be problematic. Resilience will be hard to identify. Those who are left will be looking for different words.

It means we should focus more than we have been on the set of disasters—very small, rather unlikely in any given time interval, but inevitable over long enough periods—that would be showstoppers for the human race as a whole, given today's resiliency.

These include but are not limited to an impact from an asteroid perhaps five miles across, something the size of the K-T meteor that may have caused the extinction of the dinosaurs 65 million years ago; a pandemic such as the black death, which killed a third of the people from India to Iceland over a single winter in 1347–1348; a nuclear war; and climate change. Over time, there may be cyber threats that might assume such impact.

Nations of the world are investigating many such scenarios, exploring options for prevention and mitigation. That said, the level of investment is small. We might consider doing more, both domestically and in concert with other nations.

Real-World Living Initiative #4. Institute a basket of hazards policies, stressing (1) no-adverse impact; (2) learning from experience and forming an independent National Disaster Reduction Board toward that end; (3) keeping score; (4) building public–private partnerships; (5) recognizing and reframing part of the Department of Commerce portfolio to achieve these goals; and (6) establishing pilot programs to identify and forestall global threats.

8.8. Recap

Even the most preliminary exploration of just one element of the policy landscape reveals myriad opportunities for improving the human prospect— in this case with respect to reducing the loss of life, property damage, and economic disruption from natural hazards.

Some features of this landscape are more accessible than others. Progress in reducing hazards losses has been uneven and will probably continue as such.

Progress is *likely* over the next decade with respect to advances in the following[9]:

1. *Warning and emergency response.* If anything, the progress here has far exceeded what I'd expected only 10 years ago. The front-end part—the observations and the modeling—has made great strides, as evidenced by the Hurricane Sandy forecast. Emergency responders have grown more savvy in their use of forecasts and in prepositioning assets prior to any emergency. And social scientists are focusing on the content of warning messages.

2. *Insurance, other mechanisms for spreading risk.* Hurricane Andrew, 9/11, and the financial-sector meltdown of 2008 have each triggered the development of new financial mechanisms for spreading risk over ever-larger pools of assets. (However, the sector's tools are limited. Despite attainments in *spreading* risk, it has yet to succeed in markedly *reducing* risk.)

3. *Information access.* There was a time when disaster zones were also information-starved. Now, thanks to IT and social media, the disaster zone is information-rich, even when and where cell phone infrastructure is temporarily saturated. Search and rescue and many other elements of emergency response are being transformed in consequence.

---

9. Much of the material in this section is taken from a post on LivingontheReal-World, dated May 15, 2013.

Progress is *possible but more problematic* with respect to advances in these areas:

1. *Public awareness and education.* The media have discovered that public interest in catastrophe is essentially unlimited. A whole lot of learning is going on every evening in front of the television and the computer screen. Schoolchildren are utterly fascinated by weather extremes, and this continues to be a gateway for American children into science more broadly. But knowledge in the abstract doesn't always translate to more effective behavior in the actual hazard event.

2. *Pre-event mitigation.* There's growing appetite for this, especially at the local level. Local officials and individual homeowners are increasingly aware that Washington bailouts after a disaster are likely to be slow in coming, less than needed, and in any event will not make them whole. Community leaders are realizing that community survival is a holistic thing. Small businesses won't survive and large business won't avoid disruption unless the workers, their families, their schools, their hospitals, and more aren't equally resilient. As for the flip side, a community isn't resilient unless the jobs are still there after the hazard has come and gone. Appetite and enthusiasm for the NOAA–NWS Weather-Ready Nation initiative reflects this.

3. *Sustained international cooperation.* Emergency response and disaster relief continue to bring countries together, if only momentarily. But progress with respect to pre-event mitigation measures continues to challenge multinational approaches.

Success may prove more elusive with respect to these principles:

1. *Reducing vulnerabilities of critical infrastructure.* There are several challenges here. Dependence on critical infrastructure (communications, power, transportation, sewage, water, healthcare, schools, or financial) is still relatively new, historically speaking. We haven't accumulated a lot of experience in seeing how critical infrastructures can fail and how disaster impacts cascade through such infrastructures to immobilize communities and even entire nations—look at the great Tohoku earthquake and tsunami and its effect on Japanese (mainly nuclear) electrical utilities. And the costs of retrofitting critical infrastructure to build in resilience and continuity are staggering.

2. *Challenges posed by megacities.* The world's megacities are essentially massive job shops competing for global business. The competition is often based on price and to keep costs down. Megacities build in floodplains and on fault zones, and compound risk by jerry-rigging cheap but fragile infrastructure. Poverty is often endemic. Hazard resilience is low.

3. *Newly emerging hazards.* It is sadly true of all of us individually and collectively that we learn best through practice. By definition, we've had no experience with newly emerging hazards. Threats waiting in the wings include pandemics, major asteroid strikes, weapons of mass destruction, cyberattacks, and more. These portend bitter future lessons.

4. *Equity issues.* This challenge is old but stubbornly resistant to cure. All evidence shows that disasters aggravate preexisting social inequities, whether based on gender, religion, ethnicity, or economic fault lines. Jesus said the poor are always with us, and he could have added that they would also always be more disadvantaged by hazards.

These last four issues are particularly challenging because they're interwoven.

Chapters 6–8 have suggested that policies are potentially a powerful tool for reconciling resource, environmental, and hazards challenges simultaneously. But today's policies are often deficient in important respects. To realize their potential therefore requires additional steps: (1) innovating; (2) quickly putting improved policies into actual practice; and (3) early on, and continually thereafter, assessing the performance of those policies relative to the outcomes desired. Meteorologists have some experience in numerical weather prediction that might suggest some ideas on how to proceed, as described in Chapter 9, which offers some initial ideas and perspective.

9

TAKING POLICIES FROM ABSTRACTION TO ACTION

Four-wheel-drive vehicles don't get hung-up any less than the two-wheel-drive kind. They just take you further off-road before you do. —*widely known and variously quoted Colorado aphorism*

9.1. The 21st-Century Human Challenge

Our defining 21st-century challenge? Balancing our efforts to manage resources, protect the environment, and build resilience to hazards. In practical terms, that translates to getting our resource, environmental, and hazards *policies* right; taking them from mere abstractions; putting them into widespread practice; continually evaluating how we're doing; and accomplishing all this "in time."[1]

Why bother trying to improve policies at all? The quote above makes the point. The power of policies lies in their emergent properties and in the speed with which they allow all seven billion of us to make decisions and take action. They're much like that four-wheel-drive truck or SUV. The

---

1. In time? Meaning while we still enjoy many years' worth of resource stocks, especially of nonrenewable resources; while we still enjoy substantial natural landscape, biomass, and biodiversity; and before the worst of the hazards of the geological record return. Note that we don't know how much time we have. There's urgency here.

horsepower under the hood and the drivetrain torque, the four-wheel traction, and the winch can take us deep in the back country before we find ourselves stranded. And it's only the last few feet of the driving, not all those early miles, that creates the problem. It usually feels like smooth sailing right up until that point. In the same way, national and global policies can get us into trouble before we know it. It's precisely the reason that the 21st century is problematic.

The majority of us spend most of our waking lives living in the moment and thinking inside the box, living and acting on the basis of policies, not unlike the ants of Chapter 6. We (mostly) all do this (most) all of the time in order to make decisions and act sufficiently rapidly to keep up with the pace of real-world events. Life is coming at us fast.

But as we saw in our discussion of the "boids" of Chapter 6, slightly different policies can lead to wildly different outcomes when broadly applied for extended periods. Here's a grossly oversimplified, exaggerated comparison between two sets of policy "options" for living on the real world. Let me emphasize: I'm not asking that you see either of the policy options presented here as better than the other. I'm not suggesting either or both are realizable in practice. This is merely an entirely notional matrix to get us thinking about the policy opportunities.

The matrix seems tame enough; some of the differences between the two options might at first blush even appear bland. But suppose, for the sake of argument, that throughout the rest of the 21st century countries and institutions frame the three issues individually and as a group according to Option 1. They see resources as limited. They see their individual or national role to capture, control, and perhaps squirrel away as many and as much of limited resources as possible, against that day when the shortages grow more acute. They seek to take and hold as much as possible through superior financial position or through greater foresight or through armed might. (We see these tendencies operating worldwide at the moment, as countries such as China, Saudi Arabia, and others compete to lock in future supplies of food, water, energy, and nonrenewables such as iron, copper, and the like.[2]) Countries and companies in the process of acquiring resources damage the environment inadvertently but then later try to restore habitat and endangered species. They could try to engineer hazards away with building codes, levees, dams, and the like—approaches that work only up to some limiting

---

2. Some analysts have also seen U.S.–Mideastern foreign policy as shaped similarly, stemming from a goal to protect and stabilize U.S. access to petroleum.

TABLE 9.1. Policy Options for Living on the Real World.

Issue	Policy Option 1	Policy Option 2
resources	acquire and consume	create and share
environment	first damage/then restore	preserve going along
hazards	build resistance to extremes	build resilience to extremes
resources, environment, hazards	"solve" each in isolation; seek artificial exactitude	"solve" all three simultaneously but settle for "approximately"
(all the above, taken together)	compete with others; collaborate in experiments; repeat mistakes	collaborate with others; compete in experiments; learn from experience and innovate

threshold (wind speed, flood stage, seismic strength, etc.). Across the board, they could see the world as zero sum, requiring that they compete, except possibly for occasional, short-term, cooperative research ventures (international space stations, high-energy particle accelerators, Antarctic studies, etc.). Learning is slow. There's a regrettable tendency to favor business as usual and make the same mistakes over and over again in new contexts.

What would you forecast such a world to look like in 2100?

Chances are good you would foresee a world characterized by water, food, and energy shortages, shortages that would be particularly acute at some places and times. You'd likely anticipate a degraded environment or loss of biodiversity. You'd see losses to natural hazards continuing to rise as rare but cataclysmic extremes exceeded the design parameters of the built environment and critical infrastructure. You'd expect that the competition would lead to winners and losers, displaced populations of the poor—the losers of such competition. You'd brace yourself for a backdraft of terrorism and war requiring constant repression to keep things under control.

This is the apocalyptic future about which many experts write.[3]

Alternatively, countries, institutions, and individuals could try a framing over the next 100 years that looked a bit more like that of Option 2. They could emphasize innovation and the *creation* of resources.[4] (The IT revolu-

---

3. Futures anticipated, for example, in Raskin et al., *Great Transition: The Promise and Lure of the Times Ahead*, published in 2002 by the Stockholm Environment Institute: specifically, the two (out of six) scenarios they labeled Barbarization and Fortress World, respectively.

4. Or (not quite equivalently) a reduction in the natural-resource intensity of their economies: that is, the extent to which their economies are built on natural resource extraction.

tion, biotechnology, nanotechnology, and other advances come to mind.) They could attempt to improve their foresight, their ability to spot incipient environmental threats from a long way off, and *preclude* them, head them off, versus merely *fix* them. (It's unlikely we can ever be entirely successful here but let's suppose that in the effort we can do better than we've been doing.) Countries could accept extremes as a recurring reality and focus less on emergency response and more on building resilience into communities in advance, so that communities bend versus break in the face of extremes. They could achieve this, for example, by focusing more on soft approaches like land use and relying less on strenuous building codes, especially along coasts, in floodplains, and on seismic zones. They could continue to compete and excel at creating resources, options, and buying time, but they would compete primarily at innovation, in hopes that diverse approaches to common goals might offer the best chance of identifying promising new ways forward.[5] More generally, however, once the research results were in, they would also see and be open to the possibility and the benefits of collaborating and sharing. And they would devotedly learn from experience and avoid to the extent possible making the same mistake twice or letting dysfunction persist.

(Not so many people are writing about this future. That might be because apocalypse sells or perhaps because to write in this way about the future seems Pollyannish.[6])

Chances are good that you recognize that the posture of your community (as defined either geographically or in terms of discipline or profession) or your nation lies somewhere between these extremes. It ought to be clear that pursuing either set of rules over time will lead to substantially different long-term outcomes. (You might try another exercise here: See if you can capture the set of rules [or values or culture or policies or mindset] under which you and your community operate.)

Here's another way of looking at the policy challenge, a way that highlights the iterative, time-step nature of the problem versus two snapshots, taken a century apart. It's captured in Figure 9.1, which is meant to indicate

5. Some have suggested that had the Greeks and the Arabs been in better communication centuries ago, we might have either plane geometry or algebra, but not both.

6. But the Pollyanna of the original story doesn't see the error of her ways and become a cynic like everyone else in her town. She transforms her town instead. Check it out. One place to start might be the post on LivingontheRealWorld, dated December 12, 2012.

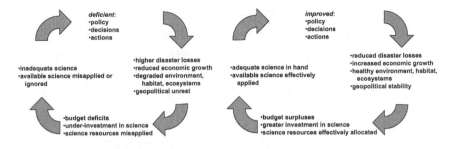

FIGURE 9.1. Two outcomes compared.

what happens in a given unit of time (which might be a year or so, for example).

The diagram works at both local and global levels, and at scales in between. The diagram at left allows for the possibility that we might get along fine, even for a number of years, but suggests that eventually the risks inherent in depending on flawed policies will catch up with us. The week or month or year will come that brings a flood or hurricane or earthquake or a punishing drought that will devastate the local agriculture. The resulting economic losses and environmental degradation will compromise the economy and perhaps drive government to scale back its support for

Each year, hazard losses increase. And, in consequence of all these trends, human beings grow more restive, dissatisfied, unsettled. What's worse, the more vulnerable of the world's seven billion find themselves malnourished, starving, homeless, and/or on the move.

science. With inadequate science we may be even more vulnerable to environmental risks in subsequent years. We've embarked on a vicious cycle. That's how things look to an individual, or to the residents of a region, or to employees in a specific energy sector. But when we aggregate all these losses across all seven billion of us we find that what is an occasional event locally or by sector is an ongoing process somewhere in the world with respect to some or another aspect. In the business-as-usual or persistence forecast scenario (above, left) we ratchet a little bit each year in a negative direction. It gets just a little more difficult to extract the resources we need from the earth. We lose just a little more habitat and a few more species;

the environment degrades. Each year, hazard losses increase. And, in consequence of all these trends, human beings grow more restive, dissatisfied, unsettled. What's worse, the more vulnerable of the world's seven billion find themselves malnourished, starving, homeless, and/or on the move. Moreover, the dysfunction from each year makes trying to do better the next year just that much more difficult.

Unlike ants, whose behavior represents best practices that have survived millions upon millions of trial-and-error experiments that have served to eliminate those ant colonies with even the slightest flaws, how we're faring depends on decisions and actions that we're essentially making up as we go along. We're winging it. We've got some past experience to guide us, a record of successes and failures. But we perhaps don't know enough, or perhaps our selfishness and stubbornness and shortsightedness are preventing our using effectively what we do know. In particular, except for a tiny minority, most of the world's seven billion people—and most of the local governments where they live and most of the small to midsize corporations where they live—are going through each time step with little or no consideration for the long-term aspects of the three challenges: resource extraction, environmental protection, and disaster resilience. For some, this is a matter of desperation; the world's poorest people can have no goal other than making it through each day. For those of us in the developed world, this is a matter of not knowing how to, even when we want to, achieve all three goals of resource extraction, environmental protection, and vulnerability to natural extremes.

Thus the diagram and the matrix above by no means represent humanity's intentional, strategic decisions about how to move forward. They are also a pale, evanescent representation of the structure and complexity of policies and the intricacy with which policies of every sort and description are woven throughout our national, institutional, and individual decision-making. That said, they may nevertheless illuminate what we're actually doing, willy-nilly.

The suggestion here is that we can become a good deal more intentional about living out this diagram and the matrix. We can grow more disciplined in keeping track of how well we're doing and making adjustments as we go along. Furthermore, the thesis is that as we improve, we just might find it possible to transition from the left's vicious circle to the right's virtuous cycle. Moving from where we find ourselves now toward the right side of the diagram is, or should be, the world's common goal.

But this confronts us with a major 21st-century challenge: Just what is involved in improving policy and in taking policy initiatives from formulation into practice? And how rapidly can this be accomplished and evaluated?

9.2. Policy Change at the National Level

Without continual growth and progress, such words as improvement, achievement, and success have no meaning. —*Benjamin Franklin*

I have a great respect for incremental improvement, and I've done that sort of thing in my life, but I've always been attracted to the more revolutionary changes. I don't know why. Because they're harder. They're much more stressful emotionally. And you usually go through a period where everybody tells you that you've completely failed. —*Steven Jobs*

Successful politics is not about finding people who agree with you. It is about making difficult decisions without killing each other. —*Anthony Zacharzewski*

The contrast between the first two quotes says something about the tempo of 21st-century living versus life's pace 250 years ago. The contrast also captures a bit of the urgency we need to bring to policy analysis and development over coming decades. By making suitable investments, we can readily accelerate our development of knowledge and understanding about how the real world works, with respect to both its natural and social dimensions, as discussed in Chapter 5. That's good news. It's also reassuring to see that in all likelihood there exist myriad policy options that might better realize the societal benefit from such advances in natural and social science, as the examples of Chapter 7 and 8 illustrate.

However, though we've seen that policy, once implemented, speeds the decision-making process, we haven't discussed the speed at which policy itself progresses. It turns out that public policy indeed evolves, but only slowly and deliberately. That's particularly true when we look at the status quo at the level of national legislatures or parliaments. Alone, this top-down, wholesale implementation of policy changes over entire nations or continents or worldwide will proceed far too slowly for our purposes. Think of policy as comprising several dozen dimensions that more or less correspond to the topics you see in the newspapers: jobs, healthcare, defense, science, education, social safety net, and so on. Legislators introduce bills constantly, but most such bills are introduced for show, to impress the constituents back home. The majority are inconsequential; legislators do no more than commemorate a special occasion or name a special day (such as designating March 14 as National Pi day, to highlight and support STEM education). They don't touch these larger issues.

Historically, modification of policies of substance in any given arena that matters might happen at most once every few years. Viewed from a historic perspective, this makes sense, for several reasons. First, it takes time to gauge a law's effectiveness with respect to its given intent. Such effectiveness is difficult to estimate a priori because, as Chapter 6 notes, the strength of policy is that its impacts and outcomes are emergent. The desired and the unintended consequences don't visibly manifest themselves immediately. Second, businesses and institutions of all sorts as well as individuals are making many multiyear choices that require policy stability over a corresponding period. Take something as simple as a company's hiring decisions. They're shaped by prevailing wages and trends, which are tied to economic conditions, international trade policies, and more. They're affected by contributions the company must make to social security and to healthcare. And so on. For individuals, the simple decision of whether to take out a mortgage and buy a home versus renting one depends on the stability of the person's job, the outlook for the economy, trends in interest rates, the mortgage deduction in the tax code, and more. And third, the number of potentially harmful policy options far exceeds the good. Many legislative arrangements have therefore been contrived more to prevent the passage of bad legislation than to foster the development of good policy. Perhaps nowhere is this more true than in the United States.

As a result, significant (as opposed to incremental) policy change often has taken longer, even decades. Consider healthcare, immigration policy, the tax code, gun control, and the war on drugs.

All this is the rosy scenario: How things work when the policy machinery has been humming, working at its best. Today what we see instead is something less than this ideal. The national-level policy arena is increasingly gridlocked. Legislators are unable or unwilling to explore change of any stripe. It seems that America is politically polarized and split nearly 50–50 on every issue. In fact, the policy process has evolved toward this state. Back in the day, America would see the occasional bill pass by an overwhelming majority. But today's politicians consider such universally appealing legislation to be inefficient. Whichever party holds the trump hand at the given moment prefers instead to load it up with riders to the point where it'll just barely pass. In such a contentious environment the potential for policy innovation and learning from experience at the national level is greatly reduced. All manner of suggestions for improving healthcare, the tax code, gun control, immigration, science and technology, trade, education, agriculture, and more have failed passage, not once but repeatedly, in recent years. That's where Mr. Zacharzewski's quote at the top of the section comes in.

9.3. Lewis Fry Richardson and Numerical Weather Prediction (NWP)

If meteorologists and policymakers would compare notes, they might see an analogy at work. Both communities might see a degree of similarity between the way *policies* function in our lives and in society and the way that meteorologists use physical laws (as captured, albeit imperfectly, by numerical approximations and mathematical equations) to describe the motion and evolution of Earth's atmosphere. Let's take a look.

Earth's atmosphere responds to forces and boundary conditions to produce a wonderfully variable weather, with times and places of sunshine and calm punctuated by violent storms and extremes. Centuries of science have produced a set of equations describing all this: the so-called Navier–Stokes equations. Figure 9.2 is one version.

This is a not a science text, and so I'm not going to define the terms you see here. (Shame on me.) Let's just say that to the meteorologist, these equations are a thing of beauty. Just to look at these equations evokes a visceral, emotional, even spiritual experience not that different from what all of us feel when we look at a Gothic cathedral or da Vinci's *Mona Lisa* or hear Handel's *Messiah* (or maybe you prefer Jimi Hendrix's rendition of America's National Anthem, as he played it at dawn at Woodstock in 1969).

All you and I need to know for present purposes is that no less than Mona Lisa's enigmatic smile, the Navier–Stokes equations are essentially impenetrable. They can be solved in closed form only for the simplest cases. When it comes to weather prediction we don't solve them so much as approximate them. And we don't treat them globally, but locally. Meteorologists break up

$$\frac{\partial \rho}{\partial t} + \nabla \cdot \rho u = 0 \qquad \text{conservation of mass}$$

$$\frac{\partial u}{\partial t} + (u \cdot \nabla)u = -\frac{1}{\rho}\nabla p + F + \frac{\mu}{\rho}\nabla^2 u \qquad \text{conservation of momentum}$$

$$\rho(\frac{\partial \varepsilon}{\partial t} + u \cdot \nabla \varepsilon) - \nabla \cdot (K_H \nabla T) + p\nabla \cdot u = 0 \quad \text{conservation of energy}$$

FIGURE 9.2. Navier–Stokes equations.

FIGURE 9.3. Slide rule, courtesy of http://babylon.acad.cai.cam.ac.uk/students/study/engineering/engineer05/ceengdes.htm.

Earth's atmosphere, globally and throughout its depth, into millions of small boxes or cells. Within each such volume they represent these equations by means of simple arithmetic formulas that will give close to the right answer for a very short time interval: a few minutes, say. Then they keep repeating the calculations for successive time intervals, on out to some time horizon, say a week or so (the results grow progressively more uncertain as this time horizon is extended).

Meteorologists have been doing this again and again for a little more than half a century, going back to Princeton and the first digital computers built around 1950. In fact, this method dates back even further to almost a full century, if we go back to a heroic Englishman Lewis Fry Richardson. Richardson tried to do this single-handedly, using only a slide rule and a table of logarithms.

(Younger readers may not know what a slide rule looks like, so I've included an image in Figure 9.3. As for the table of logarithms, it's even more gruesome. It makes the telephone directory look positively captivating by comparison. Now imagine looking at those equations above, arming yourself with this slide rule and those log tables, and setting out boldly. Not much different from David going after Goliath with nothing more than a sling and a few smooth stones. Lest you start feeling too smug, keep in mind that a decade or so from now, your current smartphone will look similarly so-yesterday. Remember clamshells?)

More accurately, Richardson attempted to make what meteorologists call a *hindcast*. We haven't discussed that notion until now. In short, a hindcast is like a forecast except that instead of initializing the calculations from the present time, the meteorologist picks a time in the past as the starting point, runs a "forecast" procedure from that past time, and then compares the result with the corresponding actual past evolution of the weather. A useful technique to be sure, especially when computing is slow and can't keep pace with the evolution of the weather itself; but it leaves something to be desired. After all, we all know that hindsight is 20/20; that is, we can find ourselves, however subconsciously or unintentionally, tuning the results. And hindsight doesn't help us avoid trouble; it only explains or labels what just hit us.

Remarkably, Richardson gave this a try during the height of World War I, within shouting distance of the front lines. A devout Christian and firmly pacifist, Richardson was serving with a Quaker ambulance unit in the north of France, daily ferrying World War I wounded away from the battlefield from 1916 to 1919. He registered as a conscientious objector, opting for this form of service even though it would bar him from holding any major English university post for the rest of his life.[7]

Richardson chose to simulate the weather of Europe prevailing on the day of May 20, 1910, starting at 7:00 a.m. (local time). He attempted to run out the calculation for six hours, to 1:00 p.m., on that day. It took him six weeks (nearly 200 times as long!), making all the calculations by hand, to achieve his simulation of that six-hour period. We're told that his hindcast was a bust. He "predicted" a dramatic pressure increase that failed to occur. (Scientists since have reexamined his equations. They tell us that had he applied certain principles of today's techniques for smoothing his initial data, he would have gotten a realistic result.)

By this measure, Richardson's first effort didn't go well. But he was undeterred. He had a vision, which he subsequently published in a 1922 paper titled "Weather Prediction by Numerical Process":

> After so much hard reasoning, may one play with a fantasy? Imagine a large hall like a theatre, except that the circles and galleries go right round through the space usually occupied by the stage. The walls of this chamber are painted to form a map of the globe. The ceiling represents the north polar regions, England is in the gallery, the tropics in the upper circle, Australia on the dress circle and the Antarctic in the pit.
>
> A myriad computers[8] are at work upon the weather of the part of the map where each sits, but each computer attends only to one equation or part of an equation. The work of each region is coordinated by an official of higher rank. Numerous little "night signs" display the instantaneous values so that neighbouring computers can read them. Each number is thus displayed in

---

7. Julian C. R. Hunt provides a readable and thorough biography of Lewis Fry Richardson. You can find it in the 1998 *Ann. Fluid Mech.*, **30**, xiii–xxxvi. It's also available online at http://www.cpom.org/people/jcrh/AnnRevFluMech(30)LFR.pdf.

8. In Richardson's time, a "computer" was a human being who performed mathematical calculations.

three adjacent zones so as to maintain communication to the North and South on the map.

From the floor of the pit a tall pillar rises to half the height of the hall. It carries a large pulpit on its top. In this sits the man in charge of the whole theatre; he is surrounded by several assistants and messengers. One of his duties is to maintain a uniform speed of progress in all parts of the globe. In this respect he is like the conductor of an orchestra in which the instruments are slide-rules and calculating machines. But instead of waving a baton he turns a beam of rosy light upon any region that is running ahead of the rest, and a beam of blue light upon those who are behindhand.

Four senior clerks in the central pulpit are collecting the future weather as fast as it is being computed, and despatching [sic] it by pneumatic carrier to a quiet room. There it will be coded and telephoned to the radio transmitting station. Messengers carry piles of used computing forms down to a storehouse in the cellar.

In a neighbouring building there is a research department, where they invent improvements. But there is much experimenting on a small scale before any change is made in the complex routine of the computing theatre. In a basement an enthusiast is observing eddies in the liquid lining of a huge spinning bowl, but so far the arithmetic proves the better way. In another building are all the usual financial, correspondence and administrative offices. Outside are playing fields, houses, mountains and lakes, for it was thought that those who compute the weather should breathe of it freely.

We're told he pictured as many as 60,000 people making the requisite calculations. Of course, Richardson's vision of that theater of human computers never came to be. In the 1950s, people substituted electronic computers for Richardson's human ones, beginning with the ENIAC and other similar devices.

Here we arrive at the start of a beautiful part of the story, loaded with potential for our 21st-century challenges. Slow as the ENIAC and those early computers were, they were still faster than any reasonable collection of human counterparts would ever be. Moreover, they were indefatigable, error-free, and destined to get better/faster/cheaper with the passage of time. Since then, thanks to a doubling of computer speeds nearly every 18 months and thanks to continuing attention to problems arising from the fact that the numerics necessarily deal with finite cells and time periods, *numerical weather prediction* (NWP) has similarly grown increasingly detailed and capable. To start, the cells or boxes representing the atmosphere were necessarily rather chunky (many tens of kilometers on a side), the time steps were relatively large, and the arithmetic calculations approximated only crude representations of the actual equations. But increasing computer speeds (up to 20 trillion calculations each second) have made it possible to decrease the cell size (now down to a few kilometers even for some global models), shorten the time steps, and reduce the number of fudge factors initially needed to make things look right. Forecasts are getting better in corresponding measure. Today a seven-day forecast is better than the early one-day forecasts used to be.

9.4. NWP and Living on the Real World

As a result of this half-century of experience, meteorologists are positioned to see similarities between NWP and humanity's approach to the business of *living on the real world*. For example, they might associate the three tasks of (1) developing and maintaining water, food, and energy supply; (2) protecting the environment and ecosystem services; and (3) building resilience to hazards to the *variables* (pressure, density, velocity) that are sprinkled throughout the equations of motion for the atmosphere.

They would also be quick to recognize another similarity. Although the three tasks are ultimately global and long-term, they are in practice being accomplished "locally." Specifically, in every town and region around the world, the inhabitants are daily doing the best they can to meet their resource needs; to protect habitat, biodiversity, and ecosystems; and to build resilience to hazards at a community level, on a daily basis. They may operate under and be constrained by national-level policies, but they are working toward more place-based goals. In the same way, weather develops locally around the world, at least for short time intervals. Weather a continent away will eventually influence the local conditions, but not for some period of time. At the local level, the decision makers and major actors are making most of their choices and actions in conformance with policies, not inventing a new thought process each time a question comes up. Those policies come from a variety of levels. They may be imposed by the nation, a state, or a local community. Meteorologists would also note that the three tasks are not only being addressed locally but "in the moment" or, iteratively, over a series of moments.

Meteorologists might also see a correspondence between the resource, environmental, and hazards policies governing local decisions for living on the real world and the *arithmetic equations* in NWP—the relationships among the physical parameters. The equations allow meteorologists to predict how conditions will evolve in each location throughout the atmosphere, at least for short periods of time. In the same way analyzing the policies governing the decisions and actions we're taking locally might allow us to make similar predictions about real-world living conditions. We could perhaps anticipate whether food, water, and energy might become more abundant or scarce in a specific location or region; or whether the environment would degrade or air and water quality might improve; or whether the region was growing more vulnerable to hazards or less so. Meteorologists might also be optimistic that if we summed up, or aggregated in some way, what was going on everywhere, locally, we could cobble together a global picture of current conditions and an outlook for the future.

Meteorologists could also point out where the analogy breaks down. To start, in NWP, "local" is truly local in the strict geographic sense. In the case of resource, environmental, and hazards decisions and actions, "local" may refer to a small community of practice: experts in flood mitigation, say, or tomato farmers, or tropical ecologists, or any of a myriad such groupings who may be collaborating over extended geographical distances.

The analogy breaks down in other ways. For example, in NWP the equations are placeholders for physical laws and the equations, generally speaking, are well-known and robust. They've been found consistent with observation and experiment, and they have been found to lead to other verifiable physics. Changing the mathematics would not change the weather; it would change only the accuracy of any prediction. By contrast, in the policy context, the policies are not universally agreed-upon; they vary from location to location, and they're changing with time. In the case of the weather, the equations provide a close approximation to the *simultaneous solution for all the atmospheric variables.* But in the normal policy setting, living on the real world is based on policies that treat each of the three problems (the resources, the environment, and hazards) in isolation.

In further contrast to mathematics, which is a proxy for real-world forces, policies are themselves the actual drivers in important respects. *That means that if the policies are changed, the outcomes can and will be changed.*

There are limits to this. Suppose, for example, the United States was to make it a national policy that everyone should walk on air. The result would not be the creation of a new species of air-walkers. It would be the creation of a large criminal class. We'd become instead a nation of lawbreakers. Policy formulation works best or perhaps, more accurately, works *only* when it respects physical and social realities of the type discussed in Chapters 2 and 3. The walk-on-air example may seem fanciful, but consider the drug policies and immigration policies discussed in Chapter 6. In each case, U.S. policies conflict with social realities; they therefore create social dysfunction and populate our jails.

But, apart from such limitations, changing the policies changes the actual outcomes. And the properties of policies are emergent. It's not self-evident from the statement of most policies what their end effects, including all their unintended consequences, will be. These are instead discovered after some time. The mathematical equations for NWP are something of a given. The optimal policies for real-world living are still being sought.

9.5. Recap

Most of our real-world decisions and actions with respect to resource development and use, protection of the environment and ecosystems, and building resilience to hazards occur at the local level, and for the short-term, and are based on prevailing policies rather than any new, fresh look at the problem's fundamentals.

For all of the homage we pay to thinking outside the box and thinking long-term, these are the exceptions. *Most of the time, most of us are doing quite the opposite. We are living in the moment and thinking inside the box.* It's therefore worth devoting some thought to how we can more effectively respond as a society to the big problems we face and the opportunities we might seize, given that we're not more aware.

The fact is that very little of what we do is worldwide in nature. People— all of us—think and make decisions and take actions primarily as individuals or small groups. The small group might be a married couple, family, church community, small company, or division of a big company. It might be a local government or a state government or a nation.

Again, think of these small groups as corresponding to the cells of the atmosphere used in NWP. Think of the time frame of those decisions (the next hour or the next week at work, the due date of the next quarterly report, or the baby's due date) as the time step. There are some differences, to be sure. As noted, the cells in NWP are purely spatial. The cells in our daily lives, or the divisions of our company or government agency, are more abstract and multidimensional. You and I can be located in and are limited to one spatial cell at any given moment, but most of us exist in multiple cells simultaneously for purposes of discussion here. In a given day, we make a number of decisions in the context of our workplace, others in a personal or family setting, and still others in a broad range of social networks. In today's connected society, many of those with whom we share such "cells" may be a continent or half a world away.

The reality is that most of the time we discover significant worldwide events, trends, and impacts only belatedly. That's because these are primarily the aggregated impact of all those smaller-scale thoughts, decisions, and actions. Only the passage of time gives opportunity for the analysis and reflection needed to detect and identify these, let alone distinguish between passing fad and fundamental sea change. There are things we're doing worldwide, but they're being accomplished with minimal actual coordination. And we're aware of them largely through history's rear-view mirror.

This reality would seem to contain elements of challenge and reasons for hope.

Let's turn first to the challenges. The biggest challenge is that thinking inside the box is beneficial only when we're thinking about the *right* things and thinking about those right things in the right way.

That's pretty much the case for the two "success" stories we've given. In the first instance, myriad ant colonies over millennia, even millions of years, have probably evinced numerous variations on whatever coding is operative in ants today, but all variations that proved to be problematic, had flaws, or incorporated vulnerabilities have been eliminated from the gene pool by natural selection, leaving only the variants that happen to work well. If a given ant hill physically collapsed—cutting off the oxygen or food or water supply to the ants within and exposing them to predators—then they'd preferentially die out and their faulty coding would die out with them.

The second success story, NWP, is slightly different. But it shares the feature of having a basic framework of equations that works pretty well, well enough that with sustained care and feeding and adjustments around the edges of that framework and with continuous emphasis on innovation, scientists have over the course of a century achieved an ability to predict how the atmosphere will evolve over the next several days.

Now when it comes to the routine matters of life, which occupy most of our attention inside the box, we're thinking similarly, in the right way. If our short-term objective is to buy groceries for the week, we do what we've done dozens of times before and every step of the way: from the drive to the store to the strategy for working the cart through that store to get what we need, to navigating through the shortest checkout line, to taking the shortcut home. We're almost on automatic pilot. The thought process and resultant actions are highly efficient. The same idea applies to preparing for an annual meeting, developing a week of lessons for a high school class that you've taught a dozen times, executing a big field project, or rehearsing for the band concert.

Where and when this happy circumstance breaks down is after a move from one city to another (or one country to another), or a change of jobs, or the end of an important relationship. Tasks that had become straightforward have to be relearned.

Suppose now we move from these quotidian, familiar concerns to our real-world challenge and its three components: extracting natural resources (especially food, water, and fiber); protecting the environment and ecosystems; and building resilience to hazards. Suppose we see our problem as that of solving all three of these relationships simultaneously, just as we conserve momentum, energy, and mass in the Navier–Stokes equations in NWP. Suppose we break up that problem into hundreds or thousands (maybe even millions) of local problems.

To what degree are these considerations part and parcel of everyone's thinking?

Not so much. We are dimly aware of the ads and the legislation heralding the end of incandescent light bulbs. The auto manufacturers tout their EPA mileage figures. We might know that the runoff from our lawn fertilizer contributes to algal bloom at the coast most summers. In the developed world, hundreds, if not thousands, of these factoids run through our heads on any given day. But are they actionable principles? Most likely they're not, at least for most of us. We make that grocery run in the most time-efficient and fuel-efficient manner, given the vehicle we have, but the vehicle may have been selected because it enables that once-a-year camping trip we enjoy, or because our brother sold it to us, or because it was red. Unless we work for an environmental government agency, or a specialty firm helping agribusiness cope with environmental impacts, or a water-resource consulting firm working abroad, we may not have internalized environmental considerations into our thought process. The same goes for resource extraction. You and I are too busy meeting the daunting demands of our work in the healthcare profession, the law firm, the classroom, the shop, or the assembly line to worry much about where our food, energy, and water come from. From time to time news reports will alert us to price spikes in grains or beef, or to the fact the Chinese are withholding rare metals from the Japanese and Koreans who are manufacturing cell phones and televisions, or to the fact that this or that product was manufactured in Asian sweatshops. But often this information fails to sink in or to get converted into an actual response. And when it comes to vulnerability to natural hazards, most of us feel lucky. We rejoice each day we wake up in the San Francisco Bay area because of its climate or high-tech job opportunities, and we push out of our minds any threat of earthquake and wildfire or rain-cause landslides on unstable hillsides. Something less than one-third of Oklahomans have a basement; only 10,000 safe rooms have been sold over the last decade in the heart of Tornado Alley. Beachfront property remains some of the world's priciest real estate, despite the attendant risks.

Given our disconnect with these policy issues that matter most to us corporately, we're in need of auxiliary help, a tool or tools for helping us look ahead and anticipate the likely long-term consequences of each and all of the multiple options available for our response. Perhaps we have computing power sufficient to aggregate up all the small, spur-of-the-moment decisions we're constantly making to better see and understand their consequences early on.

9.6. A Promising Initiative

Our computers and our modeling might have reached the point where we can actually model society's evolution into the future in much the same way that we model the atmosphere now. Some attempts along these lines are currently being complicated by the University Corporation for Atmospheric Research (UCAR) here in the United States and elsewhere.

This will have multiple uses, but here are two.

First, such modeling, used to forecast, might provide an early warning of coming societal challenges. At this point, some readers might argue that we're already doing something like this, in constructing alternative emissions scenarios as starting points for Intergovernmental Panel on Climate Change (IPCC) analyses of climate change, for example. But most such modeling is built up from some macro level, versus aggregating it piece by piece. Such modeling is so important to the world's future that the approaches shouldn't be seen in terms of *either-or* but rather *both-and*.

Second, such modeling could be used to analyze policy options and test their efficacy before they're actually put into practice. It might be natural and tempting to scoff at this second suggestion, but here is a thought. There was a time when new airplane designs were introduced that intrepid test pilots were the ones to put actual prototypes through their paces, to actually take them up to see if the wings would stay on and if the plane was controllable, at great risk to their lives. We still have test pilots, but these days most of the performance characteristics of a new aircraft design have already been verified by computer modeling well before the first test pilot puts his or her life on the line. The Boeing 787 Dreamliner constitutes a recent example of this practice.[9]

Real-World Living Initiative #5. Initiate a so-called L. F. Richardson World Outlook Project *worldwide, with the deliberate goal of modeling world outcomes (and downscaling to nations and regions) and their sensitivity to policy options with respect to resource use, environmental protection, hazard resilience, and more.*[10]

The fact remains that in order to reengineer our policies in time to navigate the 21st century will require that we greatly accelerate the process. Simulating the likely effects of policy options on computers, in the manner of Initiative #5, may speed things up a bit, but not nearly enough. We need some additional means of exploring the implications of myriad policy alternatives, in parallel and simultaneously rather than serially. Social networking may offer a powerful way to accelerate policy improvement by orders of magnitude, while minimizing any risk of destabilizing society. That is the subject we turn to next.

---

9. With regard to such modeling of social evolution, you have the Bill Hooke guarantee: The early efforts will be no worse than L. F. Richardson's initial numerical weather forecast.

10. In this way, aim to extend the pioneering work of Gus Speth and *The Global 2000 Report to the President: Entering the Twenty-First Century* (Penguin Books, 1982). Specifically extend that analysis to the regime he and his effort acknowledged they stopped short of handling: the nonlinear interactions of all the separate components to resource consumption.

10

SOCIAL NETWORKING

10.1. Community

Another name for social network is community. The words *community* and *communication* share a common root for a reason. Communication works best when all parties are in community. And meteorologists are a rare breed in this respect.

When I was in junior high school in Pittsburgh my science class spent an entire semester studying weather. More precisely, such was the state of the local education at the time that we spent the entire semester on weather superstitions. You know the drill: "Red sky at night, sailors' delight; red sky at morning, sailors take warning" and "Mare's tails (cirrus clouds) and mackerel scales (cirro-cumulus) make loft ships carry low sails." My teacher had no science training and was nervous about tackling the real meat of the material.[1] But there was an opening line in our science book that captivated me. It said something to the effect that scientists were "a community of scholars,

---

1. She needed today's AMS Education Program, the best (legal) Ponzi scheme extant and the best approach to equipping teachers of K–12 Earth sciences in my experience. A small staff of dedicated, knowledgeable AMS professionals work with a cadre of teachers, who in their turn each work with another cohort and so on. In this way the program has reached 100,000 teachers and millions of students.

engaged in a common search for knowledge." Because I am the son of a mathematician, have an uncle who's a plasma physicist, and with two great grandparents on that side of the family who had been a city engineer and an astronomer, respectively, this idea seemed magical. I saw my life's path. I felt I belonged and that what I was doing was important.

Fast-forward to my first year of graduate school, studying physics at the University of Chicago in 1964–1965. I had a research assistantship at the Institute for the Study of Metals (ISM). I spent my afternoons banging Unistrut into the walls and building vacuum systems. We needed the latter to grow crystals that were sufficiently pure and free from dislocations so that we could study the so-called de Haas-van Alphen (dHvA) effect, a quantum mechanical effect in which the magnetic moment of a pure metal crystal oscillates as the magnetic field is increased. Over time, this effect has become a useful diagnostic in what is today called condensed-matter physics. Back then, the nomenclature was different; I told friends I was doing work in low-temperature solid-state. I hoped someday to work in a semiconductor laboratory; that and electronics were my first loves.

But ISM back then, at least what I was able to see of it, was a sweatshop. dHvA was already a relatively well-picked-over field. The scramble was less about pushing back the frontiers of science and more about growing crystals of new metals that had yet to be run through the process—as much a metallurgical art as a science. One day a faculty member came in and said, "It took me 18 months to grow this crystal. I'm damned if I'll show anybody how I did it until I've milked it for all it's worth."

The alarm bells went off. What had happened to the community of scholars, engaged in a common search for knowledge? After one quarter, I thought I should consider a change. I started looking around campus for something else to do. I had a friend from my Swarthmore days who was in an interdisciplinary program at Chicago on mathematical biology. They needed to know some fluid dynamics, so they'd understand how blood goes through veins and capillaries. He was taking a course in geophysical fluid dynamics from George Platzman (not exactly the same fluids regime) and suggested I talk with him. Professor Platzman proved hugely helpful. He came back in three weeks, saying two faculty members had support to offer. One, Roscoe Braham, did cloud physics. The other, Colin Hines, studied ionospheric wave dynamics. I was familiar with clouds (who isn't?), but the upper air, almost as good a vacuum as I was building at ISM, sounded a little esoteric. I opted for the former. I thought it would be interesting to work on something I could see every day. It all seemed set until Platzman got back to me a little later saying that possibility had dried up. I wanted out of ISM. So I developed an instant zeal for ionospheric wave dynamics.

And walked into the sunshine.

This proved to be one of the happiest accidents in my life. I'd failed to do any due diligence, but as this book's theme suggests, meteorologists and geophysical scientists more broadly are embarked on an inherently collaborative endeavor. And they've been helped over the years by three additional realities: There are more than enough big science problems to go around; nobody is going to get rich; and nobody is going to win a Nobel prize.

I had stumbled onto my community of scholars, engaged in a common search for knowledge.

The first departmental seminar I attended showed me the difference. But first, I have to discuss a pair of Physics Departmental Seminars for comparison. On two consecutive weeks, Subrahmanyan Chandrasekhar (then on the Chicago faculty) and Tommy Gold (from Cornell) had worked through competing theories of quasars. The rooms were packed both weeks. At the conclusion of Gold's talk, Chandrasekhar stood to ask the first question, "Surely you would agree that . . . ," he asked. I forget exactly how he finished the sentence, but it was something simple, like 2+2 = 4. Of course Gold would answer "yes." There was a very long silence. The two men looked at each other, unblinking. Finally Gold answered, "No." He'd recognized this as the first step down Chandrasekhar's path and decided to take his stand earlier rather than later.

It was a great encounter; I was privileged to be in the room. But the two seminars proved to be less about getting at the truth and more about personal branding.

By contrast, the first seminar I attended in geophysical sciences was delivered by a paleo-biologist. He'd been studying some early reptile life form (can't recall the species), but the fossil skeleton showed it to be massively built, low to the ground. He conjectured that the reptile lived off tubers, using its powerful foreclaws to root these out of the soil. "Then," he said, "I consulted my paleo-botanist colleague, who came back to tell me that there were no tubers at that time and place. Conifers were the only vegetation. So, I revised my theory. I decided that the reptile leaned up against the conifers on his hind legs and used his stubby forelegs and claws to pull pine cones down from the trees and eat them. That worked well for me until I calculated the food value of the pine cones and how many the reptile would have to eat per day to meet its needs. It came to some 2,000 pounds per day. . . ." As I recall, we all left the seminar with no good answer as to how and why this creature was ideally suited to its surroundings.

Talk about uncertainties.

I soon received another dividend. As before, I was being supported on a research assistantship. But I hadn't been tasked. My predecessor had done a boatload of computer programming. After a couple of weeks, I screwed up the courage to talk with Professor Hines. "Look," he said. "I'm happy to support you, but I don't want to spend my time thinking of things to keep you occupied."

So I started learning a little geophysical science from people eager to share. They weren't withholding information until they'd milked it for all it was worth. Instead, they recognized it was worth nothing until they shared it.

10.2. Social Networking

The world's reaction to social networking largely ranges from bemusement to irritation. To many people, the networking is purely social, as in *pertaining to, devoted to, or characterized by friendly companionship or relations* and as opposed to anything more purposeful: "Yo! I'm at Starbucks! Anyone else nearby?" To some at the other end of the spectrum, social networking is actually *anti*social. It offers an impoverished substitute for true face-to-face social engagement, and in the process, because it's easier and addictive, it is trivializing, even sabotaging, our ability to relate meaningfully to one another to live in community.

By contrast, a small but growing minority is looking at social networking as a means for problem-solving: crowdsourcing.

The term merits a little explanation. One definition states it this way: "the practice of obtaining needed services, ideas, or content by soliciting contributions from a large group of people, and especially from an online community, rather than from traditional employees or suppliers." The work often satisfies two conditions: It's tedious and it can be subdivided. Though it can be conducted offline, online technology has enlarged the possibilities. The work typically draws upon self-identified volunteers or part-time workers, each contributing small bits. It's a societal response to tackling opportunities and problems created by Big Data. The astronomical community has drawn on such crowdsourcing. In the Galaxy Zoo project, for example, some 150,000 volunteers helped classify 50,000,000 galaxies. The Search for Extra-Terrestrial Intelligence (SETI) also used volunteers in its efforts. Crowdsourcing has been applied to climate change, transcribing Egyptian hieroglyphs, and more.

Not surprisingly, for-profit Internet companies are getting into the act.

Amazon's Mechanical Turk[2] is one such example. As part of their Web Services, Amazon created what they call a crowdsourcing Internet marketplace to coordinate human thought to tasks that exceed the capabilities of today's computers. The customers pose problems (human-intelligence tasks [HITS]). The workers (or *Turkers*) browse among the tasks on offer, looking for those corresponding to their skillsets or interests. There's a complex matchmaking-acceptance-payment process.

10.3. Real-World Living's Need for Social Networking

With this background, let's reexamine our fundamental living-on-the-real-world challenge. We have to make decisions and take effective action with respect to uncountable millions of tasks: developing food, water, and energy resources; protecting the environment and ecosystem services; and building resilience to hazards at the community level. Some of these tasks are simple, but most are complicated. They're coming at us in a jumble. Time is at a premium. The job exceeds our individual mental capacities. We're hustling to build our factual understanding of this real world, along the lines described in Chapter 5, and to develop policies that will speed our decision-making and actions and at the same time make them more effective and put them into practice.

But the policies we're using aren't quite right. What's worse, formulating new ones and testing them is a matter of trial

The top-down policy process is necessarily cumbersome, slow, cautious, and, frankly, prone to dysfunction. We have to accelerate the learning process, while keeping the risks attendant upon any innovation low.

and error, inherently contentious (because at the national level the stakes are so high), and in consequence very time consuming. The top-down policy process is necessarily cumbersome, slow, cautious, and, frankly, prone to dysfunction. We have to accelerate the learning process, while keeping the risks attendant upon any innovation low.

How, then, to speed the rate at which we sift through the possible policy options to find those that work? How to test and then identify the best—or broad classes of the best—that might merit the attention of nations and the world, without risking massive destabilization of our society?

---

2. Referring to the 18th-century to 19th-century mechanical Turk billed as an automaton chess player but really an elaborate hoax housing a man inside.

The answer may lie in a second piece to the American policy process, a piece that turns out to be a vibrant laboratory for trying and testing new ideas. The United States comprises 50 individual states, and these have proved to be engines of legislative innovation on just about every dimension of government's role: welfare reform, emissions legislation, energy policy, immigration, education, healthcare, you name it. Take climate change. California, Hawaii, and Minnesota are legislating economy-wide reductions in greenhouse gas emissions; in addition, West Virginia and Wisconsin require emissions reporting. Alaska, Arizona, Arkansas, Kansas, and North Carolina have established climate change commissions. Colorado, Connecticut, and Maine have climate change action plans. Just one example from one issue, but it gives the flavor.

Many countries around the world lack such a diverse policy regime internally, but again, fortunately, there's much policy-relevant innovation that goes on outside of federal and state government—at local levels, to be sure, but also within the private sector, within national government organizations (NGOs), within the university community, and within communities of practice worldwide in just about every sector of the economy or human activity. Staying with the climate change example, groups such as the World Business Council for Sustainable Development (WBCSD), Business for Innovative Climate and Energy Policy (BICEP), and myriad more all advocate in the policy arena or take company-by-company unilateral initiatives. This diversity of bottom-up approaches characterizes every policy arena. And the diversity of national approaches to issues itself offers opportunities for experimentation, observation, and study.

All this motivates us to return to Chapter 9's analogy between living on the real world and numerical weather prediction (NWP). Recall that everyone is for the most part living in the moment (focusing only on the present) and thinking inside the box (that is, "locally"). Suppose we add to that lifestyle two other dimensions: (1) learning from experience and (2) sharing what we've learned through social networking.

Note that we haven't really introduced anything truly new to our lives here.[3] Even the slowest of us learn from experience to some degree, and we're

---

3. Nor have we introduced anything that new to the literature. Many of the notions put forth here and in other chapters on local, place-based approaches to societal problems have been articulated rather more carefully (and their scientific basis documented more thoroughly) by Ronald D. Brunner and Amanda H. Lynch in their wonderful book *Adaptive Governance and Climate Change*, American Meteorological Society (2010).

all talking about what we've learned, all the time, even if what we've learned is only the latest gossip. What's being suggested here is envisioning these two functions as less of an adjunct or afterthought to life and more as the essence of it, where withholding information because it might be of short-term personal value is less of a priority than sharing to maximize and accelerate the larger societal benefit. Similarly, at the receiving end, instead of being entirely absorbed with the mistakes and experience of our own creation, we're growing slightly more open to what we can learn from the successes and failures of others, especially with respect to the real-world living pieces.

We've seen hints that operating on the basis of effective policies—much as ants construct and maintain a colony and much as meteorologists accomplish numerical weather forecasts by using arithmetic expressions that approximate the real-world physics of the atmosphere—can guide our thinking and help us make decisions and take action quickly and nimbly, with the timeliness demanded by our planet. Acting on the basis of good policy can leverage our intelligence and help us function as if we were smarter individually and collectively than we really are.

If we do this, then perhaps we might in aggregate learn in a way similar to that manifested by Martin Gardner's matchbox computer.

10.4. The Matchbox Computer

In the 1960s, I was reading *Scientific American* rather regularly. I often enjoyed Martin Gardner, who wrote a column on mathematical games for them for a quarter century. One month, his topic was a matchbox computer that could learn.

I can't tell you about any other one column he wrote. But that one stayed with me.

The notion made an impression on others as well. Fifty years later, you can still find a number of references to it on the Internet.

Here's the idea. Take a simple game, like tic-tac-toe, Nim, or hexapawn. Assemble a gaggle of matchboxes equal in number to the number of possible configurations of the game, and assign one of each of the possible configurations to a matchbox. Put counters in each matchbox corresponding to each one of the possible next "moves" for the matchbox for each one of these configurations.

Now start the game. Make a move. Then randomly choose the next move from the corresponding matchbox. Take your next turn; then find the matchbox corresponding to that new configuration of the game. Again allow that matchbox to choose one of the possible next moves.

The matchbox computer then "learns from its mistakes and successes." When the computer loses, remove the option the last matchbox chose from the set of possibilities for that particular matchbox. The computer will never again make that particular mistake to lose the game. Return all the other markers from the previous moves to the matchboxes. Keep playing, and repeat the process. If there's a winning move from a particular box, the matchbox computer quickly learns it. If instead that particular box keeps losing, eventually all the possible moves are eliminated from that box; it's empty. When that happens, soon all the moves from other configurations that lead to that one will themselves be eliminated, and so on. Eventually all the losing moves will be eliminated from the matchbox computer's repertoire. Even though every initial move was random, at the end, every move from every matchbox is a winning move. By "thinking inside the box," literally, the insensate matchbox computer has learned to play perfectly.

Of course these are quite simple games. What about more complicated games? It turns out that the number of matchboxes required grows quickly. One source I read suggested that a matchbox computer that could be taught to play the game Go in this way would be about the size of the Crab Nebula.

At this point you can be forgiven for being frustrated. Surely most of the real-world challenges we're talking about here are of this more complicated variety. Why should this process work over any realistic time span?

There are several reasons. The first is that this type of "learning" can be quite fast. For the simpler games, the matchboxes may need a trial-and-error period of no more than 20–30 games. Something similar probably occurs throughout the origin and evolution of life. Virtually no random agglomerations of atoms and molecules in the primordial soup led to anything interesting. Natural selection removes these "losing" configurations from play. But the few that do can then relatively quickly aggregate further. So the calculations of how long random molecular processes would take to generate life all turn out to be wildly pessimistic. (There's more to this, but you get the idea.)

Second, you and I may not be as bright as we like to think we are, but surely we're individually and collectively brighter than a matchbox or even a gaggle of them. Our problem is that although we think we are learning from our mistakes and changing behavior as we do in life, we tend to kid ourselves. We think we're getting a lot more physical activity than we do. And as measured by objective observers, we rather favor persistent folly as opposed to learning from experience.

Note that the matchbox computer doesn't even know the rules of the game to start. It just knows the difference between winning and losing. In the same way, approaching the resource, environmental, and hazards issues, it's not that we know the rules and refuse to obey them. We really don't know what, if anything, will work. We're trying to figure out the rules as we go along.[4]

To sum up, we need to develop and improve needed natural resource, environmental, and hazards policies. We can do this only by trial-and-error learning. Though we might get there someday, we're not quite smart enough to identify improvements by introspection or first principles. Such experimentation carried out with wholesale policies implemented in lockstep on a global scale would carry unacceptable risks and would never find acceptance. But we could reduce the risks and accelerate the learning by devolving policy formulation, regulation, and all the rest down to the local level and seeing each such locality or policy center or center of action as a pilot project. We could blur the distinction between science and practice by creating an expectation that everyone is in the business of gathering data about how things are working, sharing the results, paying attention to similar efforts underway elsewhere, and learning from that experience.

And none of this would be truly new. We would simply become more intentional and strategic about what is already underway, occurring naturally.

Real-World-Living Initiative #6. Imbed a research/learning element into many of the current de facto policy innovation efforts already underway in governments, corporations, and communities of practice at every level worldwide and accelerate the rate at which knowledge on the relations among science, policy, and outcomes is developed and shared.

Such efforts are springing up piecemeal everywhere.[5]

We can learn a few things from the example of meteorologists. Their social networking has historically taken several forms, each important to our discussion here: (1) sharing the basic data needed to make daily forecasts, and the development and sharing of research results; (2) utilizing internal M2M (meteorologist-to-meteorologist) discussions of the implications of research, prevailing weather, and NWP; and (3) communicating to the larger public.

---

4. Don't get enough of this challenge in the real world? You can try your hand at so-called discovery games, where the rules are deliberately withheld from new players because their discovery is part of the game itself, such as Army of Zero, Mao, NetHack, or Penultima.

5. One of the more interesting such initiatives is the Thriving Earth Exchange being developed by the American Geophysical Union in cooperation with others as a Centennial Grand Challenge to mark its upcoming 100th birthday.

10.5. International Sharing of Weather Data

> I remember wearin' straight leg Levis an' flannel shirts
> Even when they weren't in style
> I remember singin' with Roy Rogers at the movies
> When the West was really wild
> An' I was listenin' to the opry
> When all of my friends were diggin' rock 'n' roll
> An' rhythm 'n' blues
> I was country, when country wasn't cool.
> —*written by Kye Fleming and Dennis Morgan; recorded and made popular by Barbara Mandrell*

Meteorologists were social networking when social networking wasn't cool.

The practice dates back to *The Victorian Internet*, the title of the Standage book referenced in the Prologue. When Samuel Morse invented the telegraph, he changed the world forever, not least because he launched the formation of national meteorological services worldwide. That proliferation drove home the reality that weather was no respecter of national borders. Especially on continents like Europe populated by dozens of small countries, international sharing of data was essential to making forecasts. Thus the International Meteorological Organization was born, in 1873, arising from the first international meteorological conference, held in Brussels, Belgium, from late August to early September 1853, at the instigation of Matthew Fountaine Maury, U.S. Navy.

The history since has provided both a success story and a cautionary tale. For more than 150 years, meteorologists have shared data across national boundaries. For most of that history, the data sharing has been free and open. That is, all data have been freely shared, with the costs of such sharing shouldered primarily by the countries best able to pay. What's more, countries historically shared all their data. With each advance in instrumentation, and with each advance in information technology, this sharing has been widened and expanded.

But over the past decade or so, these time-honored agreements have begun to fray. Today some countries restrict the free data exchange to certain "basic" datasets. Full spatial and temporal resolution may not be available. Or information on esoteric parameters or from unusual sources may not be shared.

Technological innovations and social change have both contributed to these changes. For example, Europe, though occupying a land mass comparable to the United States, has some 20 national meteorological services compared with our one. The development of the European Economic Community (ECC) has led to some consolidation (witness the European Center for Medium-Range Weather Forecasting [ECMWF]), but that consolidation is far from complete, as countries have thus far decided that weather services are a national security issue requiring they retain some level of sovereignty and an independent capability. As a result, European weather services cost more than they should. Governments have required their weather services to contribute some measure of cost recovery, charging users for some specialized products and services. In the United States, events have taken a different turn. Here there's an independent weather services industry, which uses U.S. governmental weather data and the corresponding data available through the World Meteorological Organization's (WMO) international data sharing. The European government weather services thus look up from their efforts to sell certain services to find American firms competing with them, using European data they've freely obtained. This has led to European decisions to withhold certain data from international sharing.

This problem is compounded because thanks to the use of information technology in weather-sensitive sectors ranging from agriculture to energy to transportation to water-resource management, weather information is today worth more. And despite difficulties in valuation as discussed in Chapter 5, there's enough sense of value that private enterprise is finding it profitable to deploy and maintain networks of surface sensors or weather radars, for example, and sell data streams to users, including governments, as well as hardware.

To pursue these topics in greater depth goes beyond the purview of this small book, but the policy implications merit attention. And the policy implications, because they're international, are more daunting than parochial U.S. concerns. But a look at U.S. investments here surfaces a problem.

For individual scientists to learn how to be effective in the policy process—to be as disciplined in the policy process as they are with respect to their science—is daunting. In the United States, it can take years.

But U.S. policy is relatively simple. It's confined to a single culture and language and a single policy and regulatory framework. By contrast, the international policy arena is multicultural and multilingual, and it features a stunning diversity of policy approaches to common interests. To master it can take decades. Other countries recognize this and invest commensurately.

They provide ways and means for those interested to get engaged at an early stage in their careers. This is a hallmark for countries like China and Japan (it's surfacing as a problem for India, which has not made such an investment). By contrast, here in the United States, international policy for science and science for policy is an old person's game. U.S. scientists may fortuitously participate in an international field experiment at an early career stage, but to get a WMO assignment or to work with UNEP or IOC or UNESCO or any of these other policy-level groups usually comes along at the end of a career—too late to really learn the ropes.

Real-World-Living Initiative #7. Develop opportunities for U.S. early career scientists to work abroad at the science-policy interface. This could be done by some combination of the U.S. Department of State and other federal agencies, much like the Fulbright U.S. Student Program, or by private funds, such as Fulbright Scholars.

10.6. M2M Social Networking: The Map Discussion

The second lesson addresses communication within each box, each profession, each community. Meteorologists have pointed the way with a unique form of M2M discussion.

Decades ago, when most meteorologists spent a major portion of their time analyzing weather through the construction of weather maps, a tradition formed in forecast offices and in university departments—a discussion at the start of a day or forecast shift in a field office or university map room of the weather outlook for that day and its implications for the work of the office. The weather maps, and there were always several for each day and circumstance, were big and required a lot of space. The conversations would usually take place around a table where the maps were spread out. At their best, these discussions were anything but top-down, command-and-control briefings. Instead, they came closer to a friendly free-for-all, with the graduate students or interns having as much of a voice as the faculty or the meteorologist-in-charge in assessing the analysis and the forecast so long as they had some new insight or alternative view. Armed with this discussion, those involved would return to their desks and their individual duties—the aviation weather, say, or the public warnings. The services and products and advisories that they would then deliver to the larger public would have been informed by this collective judgment at the map table.

A 1991 National Weather Association (NWA) awards banquet talk by Leonard Snellman, former head of the National Weather Service (NWS)

Western Region, gives a flavor for map discussions at their best. Interestingly, Snellman harks back to his experiences at the Chicago Forecast Office, which interacted very closely with Rossby[6] and other meteorologists at the University of Chicago (in what became the Department of Geophysical Sciences):

It is an honor for me to speak to you tonight. The NWA is a forecaster's association, and I take pride in being a forecaster. My 1982 NWA Award "For outstanding contributions to operational meteorology" is a cherished honor. My talk tonight will include some reminiscing and a discussion of challenges and possible pitfalls that I see facing the NWA and forecasters in the 1990s.

I have given considerable thought to what an older, or as Tim Crum suggested "experienced," forecaster should say to the NWA. Often old forecasters hold on to the past so strongly that they impede progress and acceptance of new ideas. I have tried to avoid that pitfall by continuing to teach part-time at the University of Utah and to accept Academy of Science committee appointments. Doing these things has tested my marital bliss. My wife, Lynn, says that I have flunked retirement. Right after tonight, I plan to raise my grade from an F to at least a C. But, it is going to be tough because meteorology is such an interesting science.

To set the stage for my remarks, I would like you to recall the difference between a thermometer and a thermostat. A thermometer senses and adjusts to its environment; a thermostat first senses its environment and then adjusts the latter to the conditions for which it was set. The NWA in the 1990s has the choice of being a thermometer or a thermostat. My hope is that it will be a thermostat and a moving force in the exciting future of weather forecasting.

Some reminiscing.

You are looking at a very lucky guy who has had a rewarding and enjoyable career as a forecaster. It started in the late 1940s in the Chicago Forecast Office under Gordon Dunn, one of the best forecasters who ever worked for the Weather Bureau. I remember him telling me that forecasting then was 80 percent hard work, experience, and art, and 20 percent science, but that by the time I retired it would be 80 percent science and 20 percent hard work and experience. That forecast has verified well, just as did most of his weather forecasts.

6. Carl-Gustav Arvid Rossby (1898–1957), a Swedish-born U.S. meteorologist who studied under Vilhelm Bjerknes, served with the U.S. Weather Bureau and MIT and Woods Hole before joining the University of Chicago in 1940.

My first job under Dunn was to plot and analyze upper-air charts for use by line forecasters. Before each morning forecast was issued there was a map discussion (not a briefing, but a no-holds-barred discussion) to decide what general weather trends would guide the forecasts issued over the next 24 hours. Mr. Dunn nearly always participated. His paperwork took a back seat to weather forecasting. These map discussions were enjoyable learning experiences for me. At the time, I was an intern doing part-time graduate study at the University of Chicago. On a day when the forecast was difficult, I was being a good thermometer content in listening to the discussion until Dunn turned to me and said, "Len, what do the latest ideas on the jet stream bring to this forecast problem?" I was speechless and unprepared. You see, Dunn knew that Rossby was testing results of his new jet-stream premise and research on his graduate students, and he wanted his forecasters kept up-to-date. It is hard to believe today, but at that time there were relatively few observed winds at 30,000 feet. Also high speeds were considered anomalous. You can bet that I was well prepared for subsequent map discussions, and even became a thermostat in volunteering information on occasion.

Another important step Dunn initiated so his forecasters would be on the cutting edge of the science was to bring about collocation of the Forecast Office and University of Chicago Meteorology Department. When this took place our morning map discussions were joint affairs with professors and graduate students participating. One incident I shall never forget took place shortly after the collocation. The university people were discussing the "major" and "minor" trough and ridge movements and developments, a new concept in those days. We call these "long" and "short" waves today. The briefer stated that a "minor" trough was moving across Minnesota. He was abruptly interrupted by a forecaster almost yelling. "The hell it is. It is snowing 4 inches an hour in Minnesota." Of course both spokesmen were right.

We communicated better with the University people as time went on. If you look at the Chicago local forecast verification statistics you will see that 24- and 36-hour forecasts improved significantly during this collocation period. The University Forecast Office cooperation also resulted in publication of AMS Monograph #5, "Forecasting in Middle Latitudes," a significant publication for forecasters in the early 1950s.

It is important to note that it was middle-management leadership that motivated forecasters and gave them an environment in which they could stay up-to-date on the state of the science and incorporated as much science as possible into operational forecasting. . . .

Imagine now this kind of "map discussion" permeating the whole of human affairs, but especially the boxes and cells we're thinking of here focused on Earth's resources, vulnerabilities, and threats, in myriad permutations and combinations. Imagine daily sleeves-rolled-up discussions that bring together researchers and practitioners familiar with every aspect of the problem in the particular box and on the particular day in question. That's a means for learning from experience and for innovation.

We have the example of meteorologists to follow, and the technology on hand to make it feasible. What we lack is a culture that values and inspires such collaboration and communication, across the community "of the whole," not merely small subgroups united only by self-interest and competing instead of cooperating with one another.

10.7. Broadcast Meteorology

When it comes to communications between professionals and the public, meteorology also has much to offer.

Just as the invention of the telegraph allowed nations such as the United States to infer weather patterns and trace their movement across the country, that same technology allowed meteorologists to disseminate forecasts and warnings. Flags displayed prominently in communities across the country allowed people to see at a glance the weather in store. The invention and wide use of radio, and then television, allowed meteorologists an unprecedented reach for their information into every business and household.

Significantly, from the very beginning, (1) weather information has been made freely and publicly available to everyone and (2) this has been accomplished through public–private partnerships. In fact, broadcast weather forecasts are frequently held up as an archetypical public good. Economists use this term to describe a good or service that is both nonrivalrous and nonexclusionary. Use of a broadcast weather forecast by one party does not reduce its availability or its use by others. By contrast, hamburgers and cars are rivalrous goods. My consumption of either rules out its use by someone else. Again, once I broadcast the weather forecast, I can't really exclude its use by anyone with a radio receiver.

By way of contrast, consider the communication debacle that has gravely impaired the national discussion on climate change. Without attempting a scholarly study of *how* we got here, let's compare the major elements of climate science communication and broadcast meteorology and the public reception for each. First, climate:

- Scientists who have studied the problem are essentially unanimously convinced that (1) climate change is real; (2) it is largely exacerbated by human activities; and (3) it poses serious risks to humankind.
- The rest of the world is not nearly so convinced.
- Scientists, seeing this, are concerned that leaders and the societies they represent will fail to take needed actions (including but not limited to reductions in CO_2 emissions).
- Scientists communicate directly with the public.
- Scientists adopt the knowledge-deficit approach to communication, and, as this has failed to produce results, have become more extreme in their characterization of the problem.
- Scientists have also become policy-prescriptive in their communication.

As a result (though this characterization is unfair), climate scientists can come across as arrogant scolds. Compare this with communication of weather, especially weather threats:

- Weather forecasters who see problems coming share the findings with professional communicators—broadcast meteorologists and others—who communicate with the public on a daily basis and have built up a reservoir of trust with their publics.
- The professional communicators use language that the listening, viewing, and reading publics understand.
- They don't talk down to their audiences.
- They are descriptive rather than prescriptive. They'll warn of the approaching tornado but leave it to individuals whether they choose to head for the basement or the tornado shelter or grab the minicam and head outside.

Given the differences, it's hardly surprising that the public views climate science as controversial, while weather forecasts, though viewed as inherently uncertain, are usually accepted at face value. (In fact, the approaches to both weather and climate have shortcomings, both in the theory and in the execution. Both groups are increasingly consulting communication researchers and other social scientists about their messaging.)

10.8. Weather-Ready Nation Program
In the United States, the Weather-Ready Nation initiative advanced by the

National Oceanic and Atmospheric Administration (NOAA) NWS might provide an interesting pilot program and an opportunity to learn. Here's the background. As matters stand, the Weather-Ready Nation concept is the basis for an internal NWS strategic plan for improvement of weather services across the country. But the United States is by no means a nation of thousands of disaster-resilient communities coast to coast.

But it could be, and in relatively short order. What's needed is to equip local communities with a framework for assessing their hazards risk and developing local contingency plans for dealing with them. Something analogous was started during the 1990s under the Clinton Administration. Called Project Impact, it was the brainchild of James Lee Witt and his team at the Federal Emergency Management Agency (FEMA). They cobbled together a $25 million per year budget and allowed communities to submit proposals against those funds for developing local risk assessments and mitigation plans. The major requirement was that communities were to bring all sectors to the table: the small businesses and the larger private-sector firms; the first responders; the healthcare officials and the schools; NGOs; and so on. The program proved so popular that when the funding (quickly) ran out, communities volunteered to use their own resources, asking only that they be allowed to use the Project Impact designation. (The program was terminated in the early days of the first Bush Administration.)

Real-World-Living Initiative #8. Extend the NOAA/NWS Weather-Ready Nation initiative from an internal agency organizing principle to a United States that is truly weather ready at the community level.

Such an extension would constitute an excellent pilot project for *Initiative #5* of Chapter 9 and *Initiative #6* earlier in this chapter. It is large enough (national in scope and involving potentially 3,000 counties and thousands of communities) to provide a serious challenge yet focused on the specific problem of building resilience to hazards makes it tractable.

What's intriguing is that the American Meteorological Society, though a relatively small professional association, could be of catalytic help to the NWS in realizing the desired result. The AMS includes over 120 local chapters geographically dispersed across the United States. The AMS Education Program reaches into schools in every community. Some 1,500 of the 15,000 AMS members are broadcast meteorologists, dispersed across the United States in virtually all major markets, and able to shine a media spotlight on resilience to weather hazards as well as opportunities presented by favorable weather conditions.

10.9. Social Networking across Global Society As a Whole

It's possible with today's technology, to contemplate social networking humanity as a whole.

Real-World Living Initiative #9. Initiate and sustain a global dialog on three questions. In the late 1980s (before the term "social networking" had entered the vernacular) I was looking ahead to the year to 2000, the turn of the millennium. In a variety of settings I made this proposal and the next:

The year AD 0000 was declared retroactively. The year AD 1000 was celebrated in Europe and around the Mediterranean. But the year AD 2000 would be the first millennium truly celebrated worldwide. Everyone, whether subscribing to this calendar or a Jewish or Chinese or some other calendar, would observe the moment. It would be fitting to provide something more than an Olympics or series of national observances (or fret about the "Y2K" threat).

Why couldn't the United States make a statement to the world to the effect that for some 200 years we had stood for individual freedoms and liberties, and for democracy and it seemed at the brink of the new millennium that the world was coming our way? Why couldn't we therefore declare that as a people and a nation we wanted to dedicate the next 200 or so years in collaboration with other willing countries to make similar progress toward sustainable development? Toward that end, we could propose to start three conversations to address three questions:

1. *What kind of world is* likely *if we take no action?* At the time, this was envisioned as an international conversation of Earth scientists, with perhaps enough in common with today's IPCC process that readers can grasp the idea.
2. *What kind of world do we* want? At the time, this was envisioned as an international conversation of social scientists, scholars, spiritual leaders, and philosophers. The premise was, and remains, that as a collective we invest far too little attention to this ultimate question.
3. *What kind of world is* possible *if we act effectively?* At the time, this was envisioned as an international conversation of the world's entrepreneurs, inventors, business leaders, and others.

The suggestion was (recall, this was circa 1990) that we should host a quadrennial discussion on these three topics. Finally, as a marker to commemorate this vision, we'd establish another initiative.

Real-World Living Initiative #10. A new Smithsonian Museum. This was prior to development of the land in the city south of the Capitol Building extending down to what is now the location of the Nationals baseball stadium. My thought at the time was to convert that into more mall space, building a new Smithsonian Museum. In front of the building would be a statue of Empedocles.[7] Four Halls would encompass Earth, Water, Air, and Fire (energy). Each would treat the resources, environmental importance, and hazards associated with each medium. A biosphere in the center would cover the emergence and continually evolving role of life in general, and mankind in particular, on this planet. Millions of people would visit the museum each year and go away energized about the planet we call home and how we might contribute to living on it and caring for it. Surrounding the museum we'd put the headquarters of the relevant federal agencies: National Oceanic and Atmospheric Administration (NOAA), U.S. Geological Survey (USGS), National Aeronautics and Space Administration (NASA), and Environmental Protection Agency (EPA). Children and adults alike visiting the museum might look across the street and connect the dots, see the link between the major issues of Earth stewardship and possible careers as civil servants working for those same agencies, or the larger communities of private–sector and academic collaborators with those agencies. The idea was to open the architectural design of the museum to international competition, with the goal of creating an iconic structure/space that would instantly call to memory this grand, shared human challenge.

Of course history (so far) has gone quite a different direction. This is true in several senses. First, some of the agencies in question have since been relocated into new space. The EPA has moved from Waterfront Towers to space on Constitution Avenue and 14th Street. NOAA is now ensconced in Silver Spring. NASA headquarters occupy office space not far from the earlier site opposite the Smithsonian's Air and Space Museum. (A colleague and mentor had been kind enough to set up a dinner meeting for me with Smithsonian leadership on this concept, but in the early 1990s they were focused like lasers on what is now the Native American Museum and the Stephen Udvar Hazy Museum and had little time for this additional idea.)

Second, as the Internet has proliferated dialogue and discussion on all sorts of matters, from the smallest and least consequential to the most tran-

---

7. A Greek pre-Socratic philosopher (c490BC–430) who suggested that all matter was composed of four elements: earth, water, air, and fire.

scendent, it might be that a more regimented, IPCC-like discussion of the type proposed would be counterproductive. Social networking allows us to entertain the idea of a (lightly) structured conversation open and inviting to everyone, all seven billion of us. Nevertheless, it could be argued that just as in the case of the IPCC we're better off with one such consensus dialogue accompanied by a cloud of more innovative, freewheeling discussion threads; we could use some similar complementary structure to international conversation here.

The conversation could well prove world-changing. It might get all of us thinking a bit more consistently and creatively, in a sustained way, about living on the real world. That thinking would have positive consequences. As Abraham Lincoln famously said, "If we could first know where we are, and whither we are tending, we could then better judge what to do, and how to do it."

10.10. Recap

There are two pathways to progress and, interestingly, both are being traveled now by a few hardy pilgrims and explorers. The suggestion here is that more of us need to join in.

The first approach builds on the reality that the world's seven billion people have self-organized not just into nations and states but also into companies, universities, not-for-profits, neighborhood associations, and communities of practice of every type and description. Each one contains some element of policy formulation and examination of those policies, although perhaps not labeled or even recognized as such. In each, to varying degrees, there's learning by experience and innovation. Progress is being made.

But that progress is slow, ragged, intermittent, and uneven. It could perhaps be accelerated were all seven billion of us to become more aware of the game afoot, become more intentional about that work, and grow more disciplined in evaluating results and sharing those evaluations.

11

LEADERSHIP

... Things fall apart; the centre cannot hold ... —*William Butler Yeats*[1]

Most discussions of decision making assume that only senior executives make decisions or that only senior executives' decisions matter. This is a dangerous mistake. —*Peter Drucker*

11.1. The Age of Devolution[2]

Centralized, authoritarian models of governance have been held in high regard for so long that the term "devolution" might hold a negative connotation for some. But the implications as discussed here are actually quite positive.

Today the world's peoples are forced to confront the limitations of top-down, command-and-control approaches governing the affairs of nations and institutions of every type. The global financial-sector meltdown of 2008, the difficulties shoring up the Euro; the Chinese system of *hukou* (distinguishing between "rural" and "urban" citizens and denying equal rights to the former); and American encounters with the fiscal cliff, sequestration, and other self-generated crises constitute only a few symptoms of a far more general malaise.

---

1. From his poem "The Second Coming."
2. Another in the series of proposed names for the age we live in.

The increasingly complex nature of real-world living exceeds the intellectual grasp even of those at the top. Top-down decision-making is dysfunctional. By contrast, social networking and other social change are providing new opportunities, equipping everyone to make better decisions on the ground, close to the action. As a result of these several factors, many decisions and most of human affairs are increasingly being determined and lived out locally and in the moment. Our society looks and functions more like an ant colony or beehive every day. As for the world's leaders, they are like the queen ants or queen bees. They occupy a unique, special niche, but it's only one niche. They live in bubbles that isolate them from the world that most people live in, and they exert influence outside that bubble only in the most rudimentary of ways. In most instances, they, like the rest of us, can at best sway only those nearby and only over the short-term.

For example, pre-Christmas 2012, newly reelected President Obama watched House Speaker Boehner fail in his effort to bring Republicans in line on what to do about the fiscal cliff. The bargaining subsequently broke down.

But the trend isn't all that recent. Here's what Richard Neustadt, the famous presidential historian, tells us what some of Mr. Obama's predecessors had to say about the job:[3]

> In the early summer of 1952, before the heat of the campaign, President [Harry] Truman used to contemplate the problems of the general-become-President should [Dwight David] Eisenhower win the forthcoming election. "He'll sit here," Truman would remark (tapping his desk for emphasis), "and he'll say, 'Do this! Do that!' And nothing will happen. Poor Ike—it won't be a bit like the Army. He'll find it very frustrating."
>
> Eisenhower evidently found it so. "In the face of the continuing dissidence and disunity, the President sometimes simply exploded with exasperation," wrote Robert Donovan in comment on the early months of Eisenhower's first term. "What was the use, he demanded to know, of his trying to lead the Republican Party. . . . And this reaction was not limited to early months alone, or to his party only. "The President still feels," an Eisenhower aide remarked to me in 1958, "that when he's decided something, that ought to be the end of it . . . and when it bounces back undone or done wrong, he tends to react with shocked surprise."

---

3. From *Presidential Power and the Modern Presidents* (John Wiley & Sons, Inc., 1960).

Truman knew whereof he spoke. With "resignation" in the place of "shocked surprise," the aide's description would have fitted Truman. The former senator may have been less shocked than the former general, but he was no less subjected to that painful and repetitive experience: "Do this, do that, and nothing will happen." Long before he came to talk of Eisenhower he had put his own experience in other words: "I sit here all day trying to persuade people to do the things they ought to have sense enough to do without my persuading them. . . . That's all the powers of the President amount to."

Even presidents struggle to do better than live in the moment (the next event on a schedule someone else developed) and live in a box: in this case, the White House and that cocoon of staffers. Journalists and scholars sometimes call this particular box "the White House bubble," but the idea remains. Those people nearby, who have the president's ear, themselves have traditional leadership positions and hold traditional influence and have access to others with influence, but as soon as they begin the process of engaging more broadly, they too encounter delay, distortion, and attenuation of their messages and their intent. It's a classic filter problem. In the same way and for the same reason, their knowledge of things and events going on at the periphery of their sphere of influence is secondhand and filtered in corresponding measure.[4,5]

The most important function that these special few perform lies in a broad perspective, the articulation of values, the reaffirmation of principles, and the providing of comfort in the face of pain(and exhortation and encouragement in the face of challenge). They have more in common with pastors—including being subjected to the same sharp scrutiny of their performance and morals, being more adulated than respected, being more heard than heeded, and having little time for the contemplation and reflection that ought to be their stock in trade—than either group might like to acknowledge.

---

4. For example, as I work on later drafts, in May 2013, President Obama finds himself beset by accusations of a cover-up in his administration with respect to terrorist attacks on U.S. diplomats in Benghazi and with respect to IRS harassment of Tea Party groups applying for tax-exempt status.

5. Those interested can find these general notions pursued in more detail by Moises Naim, *The End of Power: From Boardrooms to Battlefields and Churches to States, Why Being In Charge Isn't What It Used to Be* (Basic Books, 2013).

On the real world, then, leadership remains absolutely essential yet has come to mean something different and has to be sought elsewhere. But where? With millions upon millions of place-based or sector-based boxes or centers of action, questions arise: Where do we look for leadership and how will it function in this new, devolved world? What does leadership look like? Is it still important? In what ways? How can it be effective instead of dysfunctional?

Yes, the choice is that stark. And these questions do matter. In the real world coming at us, we find ourselves challenged: Just which policy approaches (and which modifications of those) will improve our chances of simultaneously balancing and rationalizing our need for resources, our responsibility to protect the environment and ecosystems providing those resources, and our quest for safety in the face of extremes of nature? As we've seen in earlier chapters, we lack the intellect to foresee fully the consequences of our policies a priori. We're forced therefore to rely on trial and error. Any such search strategy that at a given time tests only a handful of centrally identified options will necessarily be too slow. We need trial-and-error learning simultaneously, across multiple fronts, in parallel, rather than sequentially. We can speed things up in this way provided we can and do break the world and its decisions up into myriad "boxes" or cells; push decision-making and freedom of action down to this lowest, local level; detect signs of success and failure early and quickly distinguish between the two; and share information on success and failure effectively from box to box, as suggested in Chapters 9 and 10. But there has to be some structure to this experimentation and communication. Communication has to be linked to decisions and action, to evaluation. This can most quickly be accomplished (and speed is of the essence) through locally based, universally distributed leadership.

Leadership, however, looks different in a live-in-the-moment, think-inside-the-box world.

A vignette:

A few years ago, I was at a dinner with some 15 people. Most were members of the National Academy of Sciences, serving on one of the more prestigious National Research Council (NRC) committees (yes, there is a pecking order). I was one of a handful of outsiders who'd been invited in, and I was sitting at a back bench, food in my lap. Those at the table were meeting with Thad Allen, then the Commandant of the United States Coast Guard (USCG). He was recounting the story of post–Hurricane Katrina New Orleans and why the USCG had proved more effective than the Army in response and recovery. Here's the gist (not a precise quote) of what he said:

When the Army deploys, it's only on orders from the Pentagon. The troops go out en masse. Once they have arrived, they wait for orders from the Pentagon telling them what they can and should do next, and then again, every step of the way. People from a great distance away, remote from the field, are in charge.

Coast Guard officers operate under a different set of rules. A Coast Guard officer at the scene is expected to take any and all appropriate actions that his or her resources allow . . . and when and as these run out, he or she is then to ask up the chain for more help . . . not help for what some distant group of people sees as the problem, but what he/she knows to be the problem there on the ground. . . .

Making allowances for the reality that Mr. Allen's remarks might have been just a tiny bit self-serving, for himself and for his service branch, his distinction is nonetheless sharp and useful.

He's in effect arguing that everyone in the Coast Guard is a leader. If everyone is thinking like a leader, leadership can no longer be about "being in charge." If everyone is trying to be in charge, we don't have leadership; we have anarchy. But everyone is responsible, both for seeing what's needed and for contributing to the response.

11.2. The Attributes of Local Leaders

The rest of this chapter focuses on this critical role of local leadership in a devolving world. If the world's peoples are to navigate the 21st-century task of living on the real world, the three tools we've discussed so far are necessary but not sufficient. We need a solid foundation of knowledge and understanding: a basis of facts. We also have a suite of policies to speed and guide our decisions and actions. But the policies currently in use are inadequate, deficient in one or more or sometimes several respects. For that reason, we need to sift through alternate policies swiftly and effectively. Fortunately, the world's real-world living problem largely plays out locally and in the moment. This means that we can view the human experience as millions of pilot projects on every conceivable subject being carried out simultaneously. We can use new social-networking tools to share information across the pilot programs. Learning from experience and policy improvement can be radically accelerated.

All of this will be accomplished (or not) by people—by groups but ultimately by individuals. And it will not be accomplished in time by individuals going through the motions, sleepwalking through the experience, being cautious, shirking responsibility, and/or being in denial. It will instead be achieved by people who are self-motivated, alert, not seeking risk but not being intimidated by it either—by people thinking and behaving like leaders, people with a bias toward action, changed outcomes, and accomplishment.

This is not simply a self-improvement book. A large fraction of the people reading this book are almost certainly leaders or consciously aspire to be. And the desire to lead is more universal, and the potential for leadership more widely distributed, than many people acknowledge or believe. This chapter is to help each reader get in touch with the leader who dwells inside and to recognize that leadership is not something that one develops only after being thrust into this or that widely acknowledged leadership position. Leadership is a mindset, to be adopted now, wherever one is geographically or by virtue of position. In much the same way that Chapter 7 didn't drill down to the details of policy formulation but rather gave a quick walk-through of the policy landscape, noting opportunities for improvement, this chapter hopes to stimulate readers to explore the subject of leadership more broadly, and over time, to be as disciplined in their thoughts with respect to leadership as they are with respect to their chosen professions.

For the sake of argument, let's say that leadership consists of the following dimensions:

- Having vision (seeing the world for what it is).
- Seeking first to understand, and then to be understood. (Any individual's vision is limited; we need to accumulate as many visions as we can in order to formulate our own.)
- Doing the right things versus doing things right.
- "*Being* the change you want to see in the world." —*Mahatma Gandhi*
  - Balancing invitation and challenge.
  - Being called versus driven.
  - Examining everyday life and actions.
  - Shouldering responsibility.

The following sections provide a word about each of these attributes in turn.

11.3. The Vision Thing

For years, I've used this hypothetical scenario to illustrate the fundamentals of leadership:

Imagine you find your 10-year-old son playing on the railroad tracks. You're concerned. "Get off the tracks!"

"Why, Dad?"

"There's a train coming!" Your son can barely hear your answer because the noise is deafening. He looks up and all he can see is train. The earth is shaking. Trees are swaying. He jumps off the tracks and he thanks you.

Contrast that with this scenario:

You find your 10-year-old son playing on the railroad tracks. You're concerned. "Get off the tracks!"

"Why, Dad?"

"Because I said so!" Your son looks up and down the tracks, but there's no train to be seen. He puts his ear to the tracks but feels no vibration. He looks back at you.

"Nothing personal, Dad, but come and make me!"

Most people—especially parents—would get the point immediately. But whenever someone didn't, I would expand on the idea. Leadership is not about holding power and wielding it summarily; it's just the opposite. Leaders perform their most important function when they help others notice some powerful trend or important coming, some *external* event they've overlooked because they've been preoccupied with other things.

It's all about vision.

When I started using this example, probably 40 years ago, managers and leaders quite often lived in and experienced a quite different universe from those working one or two organizational levels below. The managers were privy to briefings on larger organizational goals that had come from the top down. They had a lot of face time with their bosses. Information was sparse. It traveled slowly. It was hard to come by and closely held.

Today, this old picture is not so useful, because information is far more widely dispersed and accessible. It's not so likely that the boss or manager or leader has a superior, more-informed vision.

Some leaders admit this. Take George Herbert Walker Bush, the 41st president of the United States. His many good attributes included his unassuming nature. He was self-deprecating and perhaps no more so than when he would admit on occasion to be not being so good at "the vision thing."

The reality? He was a lot better at the vision thing than he let on. It seemed to be part of his nature to remain somewhat in the background, to shine the spotlight on others, and to invite people to underestimate him. No doubt this served him well as ambassador to the United Nations, chair of the Republican National Party, ambassador to China, director of the CIA, and vice president to Ronald Reagan—and even as father to the 43rd president.

All these are roles where too much visibility can inhibit accomplishment and limit influence.

However, *vision*, in contrast to *visibility*, is vital for leaders, even Bush Senior. A leader's credibility starts with being able to see things, with clarity, that matter to others.

"The vision thing" starts with clarity and realism, but it doesn't end there. Consider this Robert Kennedy quote (which he and his brothers would sometimes attribute to George Bernard Shaw):

"Some men see things as they are and say why? I dream things that never were and say why not?"

No getting around it. To be successful, leaders must *dream a* great *dream and* share *it*.

My first professional experience with this—the experience that drove home this notion—came early in my career, in 1970. Until that time I'd been working in what was then the Environmental Science Services Administration (ESSA), in the Ionospheric Propagation Laboratory, in Boulder. Gordon Little, the director of the Wave Propagation Laboratory, asked me to join him, just at the moment when the ESSA was being folded into a new federal agency, the National Oceanic and Atmospheric Administration (NOAA), and the Ionospheric Propagation Laboratory was being incorporated into a new National Telecommunications and Information Administration.

Mr. Little was and remains an extraordinary individual. A Brit (who became a U.S. citizen), he was born in China, the son of missionaries. He was Cambridge educated and would in time be elected to the National Academy of Engineering. His scientific and technical accomplishments over several decades were legion. However, please focus here on the remarkable step he took in 1967. He voluntarily stepped down from his position as director of ESSA's Institute for Telecommunication Sciences and Aeronomy, which comprised multiple laboratories. Why? In order to form a new, remote-sensing Wave Propagation Laboratory (WPL) within that complex.

Would you or I have done that? Probably not. On the surface it looked like a huge cut to his former considerable responsibilities. Why give all that up?

But over time the new laboratory, though small, proved to be a cornucopia of new tools for observing the oceans and atmosphere and learning their secrets. The work there spawned a range of innovative optical remote-sensing technologies, the wind profiler, the CODAR for measuring ocean currents, improvements to weather radar, development of active and passive acoustic probing techniques, important algorithms for the inversion of radiometric data, and much more.[6] In addition, WPL scientists and their instruments and sensors contributed to major field experiments worldwide. What the laboratory didn't develop in-house, it adapted and improved upon from laboratories around the world. And it gave back in equal measure. For decades, the Wave Propagation Laboratory played a seminal role in advancing the discipline of remote sensing and its applications.

What made WPL special? Gordon Little was a leader! He had a vision. His vision was that remote sensing observations, as opposed to in situ observations, would be the primary means going forward for observing the atmosphere and oceans. But his vision didn't begin and end there. The key to his considerable success as the director and founder of WPL was that he saw remote sensing not as an end in itself but as a key to understanding Earth, its oceans, and its atmosphere and, through that means, benefiting society. Most importantly, he didn't keep that dream to himself. He shared it, every day, every week, and every year, with the hundred-some staff of the laboratory and with other scientists, from every agency, every university, and every nation.

At first this part escaped me. For some time I used to think to myself that Gordon was a great boss—that he combined vision and integrity, and I'd be happy to work for someone who fostered either one. At the time I thought he had just one flaw. He was always repeating himself!

Then I realized that all of us in the laboratory (and indeed, many of our visitors) were learning a catechism. Like Gordon, we came to believe that there was nothing wrong with the world, no social ill, that couldn't be cured by more and better remote sensing. We knew and could recite the four great pillars of remote sensing (theory, technique development, applications, and technology transfer). We knew the seven great advantages of remote-sensing over in-situ techniques. We bought in! If we could have, we would have developed an "app" for microwave ovens (the term "app" didn't exist then), so that householders could disable the safety interlock, open the door, point the microwave oven out the window, get a quick Doppler-wind profile, and phone it in to a central location. Gordon not only had a dream, he also shared it.

But note that leaders need to have a *great* dream. It can't be a little dream. It can't be a shabby, self-serving dream. Instead, it should ennoble every

---

6. In fact, the laboratory was the incubator for a systems development effort to develop forecaster workstations for a round of NWS Modernization and Restructuring, which took place from 1975 to 1995.

hearer. It should elevate, inspire, and energize. Little wasn't thinking about how to get a bigger laboratory or greater personal prestige or become head of NOAA research or even of NOAA itself. He wasn't trying to use us to achieve his ends. He had turned his back on all that. He saw how to make the world a better place. He made sure we were all thinking the same way.

One August day in 1963 a preacher gave voice to just such a vision. He stood at the Lincoln Memorial, looking down on thousands of people gathered the whole length of the reflecting pool. He had a dream to share with them and with us.

So here's the bottom line for you and me. We are all tempted, every day, to think small, to be content with and settle for a small, shabby, self-serving dream. Maybe it's getting ahead. Maybe it's getting through the day. Maybe it's getting something for ourselves. Maybe it's just keeping what we have. What's worse, we're all too often tempted to keep our dreams to ourselves. What if someone stole our idea and ran with it? And they got the credit or the prize? Where would we get another idea?

Don't give in to these fears! The opposite is true. Remember that Henry David Thoreau quote from the Author's Intent, which advised, "Go confidently in the direction of your dreams. Live the life you've imagined." And share that imagination. Your very act of sharing will stimulate other ideas and visions in their train. As you give away and share your very finest ideas, even better ideas will come to you. If you instead try and hold on to your ideas, you'll find that the interior source of your ideas dry up, retreat, grow sterile.

One August day in 1963 a preacher[7] gave voice to just such a vision. He stood at the Lincoln Memorial, looking down on thousands of people gathered the whole length of the reflecting pool. He had a dream to share with them and with us. He began: "I have a dream . . ."

What kind of world would we have today if he had taken no deliberate action? A much-diminished one. What kind of world did he want? He told us in his speech that he wanted a world where we are all brothers. And what kind of world has proved possible in the years since, following a little action? Not a perfect world but certainly a much better one.

---

7. The reverend Martin Luther King, Jr.

11.4. Seek First to Understand and Then to Be Understood[8]

Back to George Herbert Walker Bush. He voiced doubts. We all have doubts about our handle on "the vision thing."

But the reality? Like former President Bush, you are more a person of vision than you let on, even to yourself. This most-positive trait isn't just within your grasp—it already lies within you.

To see why let's start with a little background on Pierre Mendes-France, the former prime minister of France, from 1954 to 1955.

Mendes-France was born in 1907. He graduated from the University of Paris with a doctorate in law and in 1928 became the youngest member of the Paris Bar Association in 1928. By 1932, he was the youngest member of the French National Assembly. By 1936 he was Secretary of State for finance. An up-and-comer! He was captured and held by the Nazis early in World War II, but he succeeded in escaping to England, where he joined de Gaulle. In 1947, he once again rejoined the National Assembly. In 1954, he became prime minister of France—and almost overnight a hero.

At the time he took office, the French found themselves fighting not one, but two colonial conflicts: in French Indochina (today Vietnam) and in Algeria. The drains on the French conscience, morale, and pocketbook were substantial. The Vietnamese Communists had just handed the French a stinging defeat, at the battle of Dien Bien Phu. The French could see where things were going, and they didn't like it. Anyway, Mendes-France quickly negotiated an armistice with Ho Chi Minh. Assembly members supported this action at about a 97 percent level. The French people, even though deeply troubled by the war, were stunned by this rapid turn of events. The main grumbling came from the Catholic Church, concerned about leaving so many Catholic Vietnamese vulnerable to Communism. Nevertheless, the relief was palpable. (The French would come to feel vindicated over the next decade as they watched the United States, with its far larger reach and resources, escalate the struggle and find itself and its aspirations severely wounded in consequence.)

However, mere months later, Mendes-France's government was toppled, and he was consigned to the sidelines.

What happened in the meantime? Why was his fall so great and swift?

---

8. Experiencing déjà vu? Wondering where you've heard this before? It's habit #5 from Stephen Covey's remarkable book *The Seven Habits of Highly Effective People*, published in 1989. All seven of Mr. Covey's habits merit your attention and study. All are relevant here. You owe it to yourself to give them a look.

Several factors played in. But here's a big one. In the interim, Monsieur Mendes-France had an epiphany. The French were consuming far too much wine! The statistics? We're told that at the time, France was spending 10 percent of its national income on liquor, dispensing a liter a day to its soldiers, supporting a bar for every 70 men, women, and children. Maybe Mendes-France had a point. Even with the touted medical benefits of a daily glass of wine, the French consumption might have seemed a bit over the top. However, though Mendes-France was nobly motivated, even you and I, looking at this at some remove with respect to both time and geography, can see that this could easily end badly.

And it did. Mendes-France went public with his concerns. And he wasn't just vocal, which would have been bad enough. He took action. He made a show of drinking milk on public occasions (modeling behavior—isn't that supposed to be good?), and he urged his government to do the same. He pushed through laws limiting time-honored French freedom to consume alcohol.

And he was gone within weeks. The French people saw their wine consumption as a nonproblem. It followed that he was not a visionary, but a threat.

What does Pierre Mendes-France tell us about leadership and vision?

Leadership is *not* about getting a group or a society to change course 180 degrees. Leadership is about recognizing a great, preexisting, universal, powerful but vague concern and articulating it, making it tangible, and giving it focus.

Each of these pieces matters.

1. *Leadership is not an about-face.* The crowd, the swarm we've been describing all along, is smarter than that. People may seem to be oblivious to a problem, but chances are they're not. You're far more likely to find that they're deeply troubled by the issue. Take the big issues of today: jobs and the economy; healthcare; the changing state of the world and our foreign policy and place within the new geopolitics; education; and, yes, the environment. What pollsters seem to find is that the environment isn't at the *top* of people's lists, but it's generally on those lists somewhere. The findings suggest that those of us working environmental issues make a mistake when we're tempted to frame more-shrill alarms about the coming dangers. People know that they must address these problems. It's just that other concerns are more pressing. As Steinbeck wrote, "How can you frighten a man whose hunger is not only in his own cramped stomach but in the wretched bellies of his children? You can't scare him—he has known a fear beyond every other."

2. *Leadership is about* recognizing *a preexisting, universal, powerful but vague concern.* And how can you and I do this? Not through introspection.

That would not only be difficult but pointless. Instead, we can *listen*. The best leaders, at every level, are those who follow Stephen Covey's advice: "[S]eek first to understand, and then to be understood."

This is the single key to getting a handle on the vision thing. And it's why I can say that *you are more a person of vision than you let on*. Chances are that you're already listening, to all sorts of people, in every kind of context. And without realizing it, you're also integrating what you hear. What concerns are only at the fringe? By contrast, what messages do you hear repeated—often in many different ways—on all sides? Increasingly, you and I are swimming in an information soup, and we're using that experience to see problems, and opportunities, for strategies and priorities as to how we'll spend our time and what we'll do.

The suggestion is that you're listening already. Keep it up! If anything, start listening more deliberately, choosing more strategically who and what you listen to and why; start being more intentional about your listening.

In today's world, with its time pressures, it's tempting to shortchange this vital step. But don't do that. Ask people more questions; delay jumping in quickly to offer your opinion. Pay more attention to their answers. Ask follow-up questions. Watch. Read. Pay attention to events, and to the media —a broad range of media. This skill of listening can be improved by practice. Try this everywhere, at home, at work, on the Metro, in stores.

Do this, and you'll become a person of vision. Over time that vision will grow more powerful, deeper, more insightful, and more useful. At first, many of the concerns you hear may seem vague. But don't worry. You'll become more fluent in sharpening and articulating the expression of those concerns. And focus on the powerful concerns, not the lesser ones. You'll get better at distinguishing these as you practice.

Covey has this to add:

Communication is the most important skill in life. You spend years learn-ing how to read and write, and years learning how to speak. But what about listening? What training have you had that enables you to listen so you really, deeply understand another human being? Probably none, right?

If you're like most people, you probably seek first to be understood; you want to get your point across. And in doing so, you may ignore the other person completely, pretend that you're listening, selectively hear only certain parts of the conversation or attentively focus on only the words being said, but miss the meaning entirely. So why does this happen? Because most people listen with the intent to reply, not to understand. You listen to yourself as you prepare in your mind what you are going to say, the questions you are going to ask, etc. You filter everything you hear through your life experiences, your frame of reference. You check what you hear against your autobiography and see how it measures up. And consequently, you decide prematurely what the other person means before he/she finishes communicating. Do any of the following sound familiar?

'Oh, I know just how you feel. I felt the same way.' 'I had that same thing happen to me.' 'Let me tell you what I did in a similar situation.'

Because you so often listen autobiographically, you tend to respond in one of four ways:

Evaluating:	You judge and then either agree or disagree.
Probing:	You ask questions from your own frame of reference.
Advising:	You give counsel, advice, and solutions to problems.
Interpreting:	You analyze others' motives and behaviors based on your own experiences.

You might be saying, 'Hey, now wait a minute. I'm just trying to relate to the person by drawing on my own experiences. Is that so bad?' In some situations, autobiographical responses may be appropriate, such as when another person specifically asks for help from your point of view or when there is already a very high level of trust in the relationship.[9]

Get the idea? Leadership isn't about *pretending* first to understand and then jumping in to be understood. Leading is going about that with integrity, as opposed to hypocrisy. Leadership is about listening so acutely and with such discernment that you can put yourself in the other's shoes.

11.5. Doing the Right Things

Management is doing things right; leadership is doing the right things.
—*Peter Drucker*

Here are three observations to illuminate this distinction between doing things right and doing the right things.

---

9. Stimulated? Intrigued? You might read the rest of Mr. Covey's book. If you've read it, you might consider re-reading it.

Example 1. Let's start with our infrastructure for extracting fossil fuels from the earth in the form of coal, oil, or natural gas and delivering energy to people in a form they can use: electricity, gasoline at the pump, and heat for their homes. To achieve today's infrastructure with its low cost, high reliability, and degree of safety for billions of users in every corner of the globe has required that millions of people from many walks of life have *done things right* over a span of hundreds of years. But it's far from clear that the overall enterprise was the *right thing to do*. That hinges very much on the next chapters of world history to be written, and they'll be written by readers of this book and many others. Will we wean ourselves from fossil fuels and limit the environmental damage? Will we be able to look back and say it was a great temporizing step to power the advance of society until something better came along? Or will we discover as we proceed forward that it proved a dead end, that it trapped us and became the best we were ever able to do?

Example 2. Historically, many people who have climbed to the top and taken on leadership positions in the command-and-control world of yesterday have done most if not all of that climbing by proving themselves good managers.

They did things right. They worked for someone who had a vision and made themselves indispensable to the execution of that vision. They made things hum. On their watch, the trains ran on time. They cleared every regulatory hurdle while scrupulously remaining within the law. They were never responsible for a mistake. They avoided making enemies. They went to management courses, and they were taught that the four functions of management were Planning-Leading-Organizing-Controlling (PLOC). (Please don't let the use of the word *leadership* in this context throw you; it begins to get at Mr. Drucker's distinction but only partially. With a bit of oversimplification, here it speaks more to an ability to *coax* people to follow versus *dictate*.) They took their management courses seriously. In their world and in their view, there were two kinds of people: managers and mavericks.

Meanwhile, a minority of leaders *created* the enterprises and activities they led versus working their way up some preexisting ladder with all its rules and guidelines. Recent examples include Jeff Bezos, Sergey Brin, Bill Gates, Steven Jobs, and Mark Zuckerberg. To read their biographies is to discover that they did hardly anything right. But they sure did the right things. They've created big chunks of the virtual world where today we all spend so much of our time.

When members of the first group make it to the top of the heap, they may find themselves ill-prepared to make the leap from management to leadership. Throughout their careers, even in relatively high-level management jobs, they've been followers. Members of the second group have been there all along. Don't get me wrong. We need both types. But they're different, and the individuals who know the difference and can deftly switch from one role to the other are rare. The ones who willingly do so are rarer still.

Example 3. Let's explore further this idea that managers and leaders are different but we need both. In section 11.3 we saw that "[l]eadership is not about getting a group or a society to change course 180 degrees. Leadership is about recognizing a great, preexisting, universal, powerful but vague concern and articulating it, making it tangible, and giving it focus." But as Peter Drucker noted, "The task of management is to make knowledge-work effective."

It's necessary but not sufficient to pick up on people's worries. The leader recognizes that the worried person wants to do something to alleviate those concerns but despairs of being able to make a difference. The worried person sees any effort to contribute leading instead only to exhaustion, defeat, frustration, and despair. That's especially the case with the complex, big-picture challenges posed by the world's resource needs, environmental degradation, and hazard threats. The leader's task is to provide a framework for meeting the challenge in question that is compelling; helps people see how and where they can plug in; and gives them the belief that *if they do so, their passion will find expression and actually benefit humanity, that their lives will be meaningful.*

However, meteorologists know that if you and I want to change the world, we can't blame any inaction on our part on the fact that we have no leaders to provide a framework and show us how. We have to become those leaders ourselves.

11.6. The Power of One

Be the change you want to see in the world. —*Mahatma Gandhi*[10]

Float like a butterfly, sting like a bee . . . His hands can't hit what his eyes can't see. Now you see me, now you don't. George [Foreman] thinks he will, but I know he won't. —*Muhammed Ali*

There are some politicians who make the weather. —*Winston Churchill*

---

10. As quoted earlier in this chapter.

Butterflies have been a favored subject for poets such as Muhammed Ali (!), lepidopterists, and philosophers. Butterflies' nature and their caterpillar origins lend themselves to metaphor. Through a series of happy accidents, the butterfly has come to hold special significance for meteorology, as we've seen repeatedly throughout this book.

We're told that in 1961 the meteorologist Edward N. Lorenz was rerunning a numerical weather prediction and that to save time he dropped the final three digits from a six-figure number he'd input to the previous calculation. The result was a completely different weather forecast. This chaotic nature of the atmosphere—this attribute of being so sensitive to details in the initial conditions as to be relatively unpredictable, on even the shortest time scales—has fundamentally shaped meteorological culture and thinking.

The result ultimately led Lorenz to publish his classic 1963 paper "Deterministic Nonperiodic Flow." What happened next is a bit clouded by urban legend. Lorenz is said to have heard from a colleague that if his theory were true, one flap of a seagull's wings might change the course of weather forever. Later, also according to Lorenz, when he failed to provide a title for a talk on the subject at a 1972 American Association for the Advancement of Science (AAAS) meeting, his colleague Phil Merilees entered a title of "Does the flap of a butterfly's wings in Brazil set off a tornado in Texas?" Through most of the remainder of Lorenz's life and career, he (and others) drew on this butterfly metaphor extensively. The *location* of the butterfly and the butterfly's *downstream impact* would change from lecture to lecture, or paper to paper, but the butterfly's notional role has remained invariant unto this day.

That same butterfly serves as a picture for our influence in the world. The view of the past, a view fostered by centralized, command-and-control leadership of the old school, was that the rest of us, those not privileged to be at the top of the heap, were without influence. We had little or no power to change events. In the 1840s Thomas Carlyle put it this way: "The history of the world is but the biography of great men." Carlyle's "great man" theory of history was attacked almost as soon as it surfaced, but some philosophers, among them Hegel, Kierkegaard, and Nietzsche, continued to toy with the idea up until World War II. The notion is still out there. In fact, most of us, when we examine our lives and circumstances, tend to see ourselves rather more as cogs in giant machinery than leaders forging our destiny. But the latter may be closer to the reality. For every Martin Luther King, Jr., there's a Rosa Parks.

Or maybe a thousand Rosa Parks. A competing view of history is that developments emerge from the sum of myriad smaller events. Think, for example, of the U.S. California gold rush begun in 1849 in response to the discovery at Sutter's Mill. We're told that California added 300,000 to its population from this event. The names of the 300,000 are largely unknown today, but their accumulated impact held great significance for California and, ultimately, the United States itself. Another example: some 16,000,000 men and women served in U.S. uniform in World War II. Their collected impact in prosecuting the war and its ultimate outcome far overshadowed any accomplishments claimed by the generals in charge. (The tank commanders who fought under General George "blood and guts" Patton were quick to say, "It was Patton's guts but our blood.")

Courtesy of meteorologists, then, you and I are invited to take heart from the transcendent influence of a single butterfly. We can see that like butterflies we are in the course of our lives pursuing our ordinary life goals and at the same time transforming the world. What's more, like the butterfly, we can achieve that not by winning over others through the force of our arguments, not by coercing others to emulate us but through nothing more than Gandhi's "being." And that's just the start. As mentioned in Chapter 9, the butterfly can't tell whether through its flapping it is creating a hurricane or suppressing one. But we are intentional and conscious. When and as we know our purpose in life, we humans can be far more powerful just by "being."

It's both simpler and more difficult to follow Gandhi's exhortation than it appears at first. On the one hand, the exhortation is best approached not by *adding on* tasks and attributes and all the rest, but by *subtracting*, by letting go of everything superfluous, everything that's *not* the change you want to see in the world. On the other hand, by following Gandhi's call, you're saying to those around you that "if you'll become more like me, the world will be a better place." Little room for falseness or hypocrisy there.

One area where this is particularly important doesn't get discussed as much as it should. Those who would be the change they want to see in the world should be very conscious of the distinction between being "called" versus being "driven." The driven person sees life as zero sum, sees leadership in terms of position versus being, and therefore sees leadership as a rivalrous good: Only one person can be the next to have the boss' job. The driven person sees little worth to the present except as it's used as a stepping-stone to the future. The ward heeler is dissatisfied with that role. He wants to be a city alderman. The alderwoman wants to be a congresswoman. The congresswoman is only thinking about becoming a senator. The senator harbors presidential aspirations. The called person is content where he or she finds him- or herself at any given moment. The called person is content to bloom where planted. The called person is also content with transition, whether upward or laterally or downward.

Part of "being" is balancing invitation and challenge. At our best, our most fulfilled, our impact on those around us is a blend of these two attributes. It seems to be hardwired into all of us to seek meaning in our lives, and where possible, to build more in. We love challenge, respond to it, and are drawn to those who challenge us, provided as discussed in section 11.5 that those who challenge us also offer a framework where we can fit in and find our work effective.

Part of "being" is balancing invitation and challenge.

The framework also has to include mentoring and support along the way, especially encouragement following setbacks. And although many people picture and therefore depict mentoring as a one-way street, with a mentor and a person being mentored, the reality is different. Any mentor will tell you, if he or she takes the time to reflect, that he or she is getting as much from these relationships as the people supposedly being "helped."

Along those lines, it's not widely appreciated that leaders are rarely problem-solvers; in fact, quite the opposite is true, for two reasons. First, a leader's primary responsibility is to coach others, to help them become better problem-solvers. They don't learn this by watching. They learn this by doing. And in the doing, they'll necessarily make mistakes and learn from them. By letting others do the problem-solving, the leader remains available after the fact to help them learn necessary lessons, see them as no more than that, learn to forgive themselves, and move on.

This means that in many instances, leadership is less about "follow me" than it is about "how can I help and even follow you?" My daughter, a social worker specializing in the special needs of very young children, was the one who acquainted me with this idea. She suggested in March of 2011 that I watch a TED video on what are called first followers titled "Leadership from a dancing guy."[11]

The point of the video is that for leaders to make a difference, they need followers and, most especially, they need that *first* follower. That first follower's influence is transformative. By the simple act of following, he or she transforms the leader from a nutcase into someone respectable and the leader's cause from a chasing of the wind into a worthwhile purpose.

---

11. The material following was largely excerpted from a post on LivingontheReal World, dated March 7, 2011.

The video's example is a remarkably simple one, and the case is compelling. However, we see the same notions at work in the more complex fabric of our working lives, in the business world, in the affairs of nations, and even in the world of research, as, for example, at NOAA's Wave Propagation Laboratory, discussed in section 11.3. In a very real sense, we were first followers.

In today's world there are a lot of good ideas floating around on the periphery and in the interstices of human affairs that merit a larger following. Let's look close to home at our own realm of the real world as resource, victim, and threat. We see every day in the papers and other media evidence that knowledge exists but it isn't being used; that politicians and scientists have chosen to fall into warring camps versus finding common ground; that it's easier to cluck our tongues at the world's dysfunction than to take a small, local action.

But we also see, from this same vantage point, bridgers, negotiators, and translators who are trying to span the divide between science and society. We see peacemakers who are urging a more reasoned dialogue among those of varying opinions in the world of politics and the world of science. Think of Climate Central, bloggers like Andy Revkin, Roger Pielke Jr., Judith Curry, and thousands of others. Do you agree with everything they say, all the time? Maybe not, maybe no more than the first follower of the video clip agreed with the leader on every dance move. But do you generally support the contributions they're making to the dialogue? Well then, you can help them along as well. Instead of just watching, get up and join the dance.

And closer to home, we see the American Meteorological Society, reaching beyond its comfort zone of technical journals and meetings to improve the application of its science and supporting technologies for societal benefit in dozens of ways. They and thousands of other nongovernmental organizations and faith-based organizations are laboring in obscurity and in penury to benefit society. You and I can support them!

Chances are good if you've read this far that you're already a first follower, giving credence and muscle to many such efforts. Kudos! Keep it up. And continue to be ever more strategic and thoughtful as you do this. Emphasize bringing along that second and third follower. For this follow-ship is really leadership of a type as well, isn't it?

The video states that while everyone may remember to celebrate the leader, there are few rewards for the first follower. But you and I know better. First, we know that any "benefits" for being a visible leader are really illusory. Fame? Its flip side is notoriety. Praise? It comes with criticism. Influence? It brings along jealousy and worse. Second, we know that the leader will always remember and be grateful for our follow-ship, whether he or she acknowledges it expressly or not. And third, we have the greatest satisfaction of all, the only one that no one can take from us—our own knowledge of the gift we gave and the role we played.

My daughter has a background in child psychology. It seems to be working in her profession, as well as on her children (and maybe even on her dad!). One of the maxims that she and her brethren follow is to avoid barking a steady stream of criticism but rather wait for your child to accidentally or deliberately do the right thing. Then seize the opportunity! Heap on the praise! Affirm! Be supportive!

The next time you see someone who is modeling behavior you admire, follow him or her a bit. Model that behavior yourself. Draw attention to it and to the leader. Kick-start that process of going viral. Be the tipping point.

Perhaps the last point that might be underscored as part of *being the change that you want to see in the world* is an opportunity and indeed responsibility we all have to follow Socrates' maxim: "The unexamined life is not worth living." If we're going to do better than the insensate butterfly, we have to be self-aware, particularly with respect to our impact intended or otherwise on others. Observe leaders, and you'll find that there's no one "right" style. It's possible to lead as a control freak, as a hale-fellow-well-met, by being aloof, by being approachable, by being a delegator, or by being a micro-manager (although each of these styles is unnecessarily extreme and carries its own advantages and risks). It's dangerous, however, to be out of touch with who you really are.

A friend of mine who taught two- and three-day management seminars was once teaching an in-house course at a small firm. All the managers and their boss were doing the workshop together. At one point they were having a general discussion about management styles and the boss said, "I have an open-door policy. My door is always open to my managers."

Something about the facial expressions on one or two of the managers led my friend to believe there might be more to this story. During a break he approached a member of the group for more background. Without a word, the manager led my friend down the hallway to the boss' office. On the open door there was a sign. It read:

"If you're too lazy or dumb to solve the problem yourself, come on in."

In the retelling, my friend exclaimed to me: "That boss was such a hypocrite! Wild horses couldn't make me go through his door!"

Finally, there's the little matter of shouldering responsibility. This challenge has ever been with us. So rather than add more words of my own, let me invite you to reflect on this quote by the great historian Edward Gibbon (1737–1794). When contemplating classic Athens, he put it this way:

"In the end, more than freedom, they wanted security. They wanted a comfortable life, and they lost it all—security, comfort, and freedom. When the Athenians finally wanted not to give to society but for society to give to them, when the freedom they wished for most was freedom from responsibility, then Athens ceased to be free and was never free again."

11.8. Recap

The devolution of decision-making, the integration of research and practice in an infinitude of contexts worldwide, and the exchange of information across these microcosms will transform leadership from its present form, better suited to the more hierarchical societies of the past, to leadership that is much more broadly distributed across society. The new leadership will be less about command-and-control authority and more about vision, integrity, and responsibility. It will be more deeply imbedded in problem-solving contexts on the ground and will show the dual characteristics of invitation and challenge. The good news is that large numbers of people are eager to shoulder the responsibilities that come with this role, but at the same time the numbers of those future leaders, at least initially, are far smaller than the world's seven billion. It's therefore possible to contemplate equipping and affirming them in relatively short order.

Oh, and if you've read this far, chances are good that you're such leadership material. Don't shirk either the responsibility or the opportunity. Get on with it!

12

REANALYSIS AND AN UPDATED FORECAST

The Revised Outlook for Humanity through the 21st Century

Journalism is the first rough draft of history. —*widely attributed to Phil Graham*[1]

It's time to consider the final aspects of what it means to think like a meteorologist, features that are important to refining any forecast for the future. We'll consider three. The first is a thought process and tool known as reanalysis. The second is downscaling, or taking global outlooks and forecasts and particularizing them to specific regions or locations. The third is characterizing uncertainties.

12.1. Reanalysis

As the quote suggests, the daily news is only the first draft of what becomes history after years of sifting through and incorporating additional facts,

---

1. In the 1963 speech where he supposedly made this statement, *Washington Post* publisher and co-owner Phil Graham actually said, "So let us today drudge on about our inescapably impossible task of providing every week a first rough draft of history that will never really be completed about a world we can never really understand. . . ." Not quite so punchy. Variations on the phrase can be found repeatedly in the *Washington Post* in earlier years, dating back to columnist Alan Barth in 1943: "News is only the first rough draft of history."

greater appreciation for context, and further reflection. In the same way, a meteorological nowcast is only a preliminary view of the prevailing weather at a given instant. Years later, looking back on that analysis from the hindsight of better knowledge and technology, meteorologists can and do return to historical datasets and give them a fresh look. They call the process and the result *reanalysis*. Meteorologists may undertake reanalysis for specific forecasts, such as the D-Day weather forecast for Normandy,[2] or to reexamine research datasets such as those compiled for the Global Atmospheric Research Program (GARP, 1967–1982), or for extended periods in order to improve our understanding of climate.

How might a reanalysis add to and build on Chapter 4's persistence forecast for the 21st century? Let's look a bit more deeply at the basic, introductory idea explored there, that the human race is on a roll.

In a short period of time, we humans have:

1. Greatly increased our numbers;
2. Increased our per capita resource consumption; and
3. Accelerated the advance of science and technology and the pace of social change.

How likely are these trends to persist throughout the 21st century? How will they look when revisited in 2100? Will they be as relevant then as they are today?

Most likely, not.

Let's start by reanalysis of the population trend itself. The world's population has nearly tripled over the past 60 years, but it's highly improbable this tendency will or can be maintained. Forecasts are uncertain, but no one is anticipating that by 2100 world population will be at 20 billion. For various reasons, experts project instead a leveling-off, probably at a figure of something like 9–10 billion, but perhaps even lower.

Similarly, the growth in per capita consumption of resources will likely slow, as the developing world catches up to the developed world's consumption, as the latter's consumption tapers off or perhaps even declines in response to improved technology, societal interest in other concerns (health-

---

2. For an interesting book on this topic, read James Fleming's *Weathering the Storm: Sverre Petterssen, the D-Day Forecast and the Rise of Modern Meteorology* (AMS, 2001). It's a fascinating biography of one of the world's best meteorologists and his most important forecast.

care, say), or the less material dimensions to well-being and quality of life. U.S. gross domestic product (GDP) will probably grow markedly,[3] but the resource intensity of GDP will most likely decline.

Unlike the first two trends, the advance of science and technology and the pace of social change could both continue to accelerate. True, there's a limit to the amount of additional attention a stabilized population can bring to science and technology, and there's likely to be a shift in emphasis throughout the 21st century to emerging opportunities in the biosciences and bio-technologies particularly as these relate to individual human health and the extension of life. In addition, the urgency of the real-world problems considered here—supplying the food, water, and energy to meet the needs of seven billion people, protecting the environment and priceless ecosystem services at the same time, and building resilience to extremes—is so great as to render it likely that these issues may well be settled or stabilized in the course of the coming century (as discussed further in the Epilogue). But mastery of new tools for science and technology will give each scientist and engineer of the future a greater reach and extend the speed with which knowledge and understanding can progress.

Institutions and individuals will be increasingly motivated to innovate. Innovation's benefits and its urgency will be more evident than ever. Everyone will understand innovation's role in generating wealth, broadening society's options, and buying time for civilization. Any suspense about the extent to which such creativity will be unleashed and in what ways lies in society's ability to make good on improving education—especially (but not solely) STEM education—for everyone. If governments and peoples can adopt policies favoring strong public education and making the opportunity widely available, all should be well. Fail in this important respect, and the world faces a less-happy future. (Even now, millions of young adults in developed economies are classified as NEETs [Not in Employment, Education or Training]. The figure in developing countries is 10 times as large, making up a quarter of the age group worldwide. The statistics stem from multiple causes, including low growth, clogged labor markets, and cumbersome regulations but also from a mismatch between education and work needs.)

These *likelihoods* (as opposed to Chapter 4's *realities*) hold implications for Chapter 4's five predictions for the 21st century: in particular, what those predictions might look like if extended to the 22nd century. One hundred

---

3. Reaching, conservatively, several quadrillion dollars per year versus today's $70 trillion.

years from now, forecasts with respect to the same five dimensions might look subtly or substantially different. By then it's likely that:

1. *Climate variability and change will be internalized into decision-making.* The current Intergovernmental Panel on Climate Change (IPCC) process or something like it will be a routine part of the world's business. The work of such assessments will go on, but the public flap attendant on the release of each new report will die down. For an analogy, think of economic figures that are released on a scheduled basis: GDP, balance-of-payments, unemployment rates, and so on. Everyone understands the methods used by governments to build these statistics. (In fact, there's outrage when governments such as Argentina fail to adhere to accepted practice.) An entire cottage industry of researchers and analysts has sprouted up on the periphery, making a living by suggesting modifications to the methodology, or alternative interpretations of the data. We experience the occasional dust-up triggered by an unexpectedly bad or good economic report during a tight political year, but otherwise the figures stand with little or no debate. If we could better divine the portents around us, we'd likely recognize the comparable development of a hive of activity around a similarly maturing IPCC process.

As for the implications of climate for food, water, and energy supply and demand, these too will be increasingly well understood and modeled.

2. *Risk management of hazards—land use, the engineering of buildings and critical infrastructure, and a host of related policy instruments—will more adequately forestall the worst effects of so-called natural catastrophes.* In addition, society should even be better prepared for other classes of hazards such as pandemics, cyber threats and accidents, terrorism and war, and financial-sector collapse. However, there is an important caveat here, as we've seen in Chapter 8. The history of natural hazards risk management has seen a trade-off; in areas as diverse as wildfire, flooding, and earthquake we have exchanged many smaller losses for fewer, larger ones (where *smaller* and *larger* refer to geographic extent and dollar amount as well as human fatalities, injuries, and suffering). In the process, we are moving toward two limits that are problematic.

The first is that some of the future disasters that will occur will produce worldwide losses. This poses an existential risk. Most of what we call "recovery" from disasters is not recovery of the affected population so much as repopulation of the affected area or region or industry by the outside, larger, unaffected population. When *no* population is unaffected, it's unclear what recovery will look like. However, what limited past experience we have with worldwide economic depression—with pandemics such as the Black Death and with World War II—suggests that such recoveries can be slow, arduous, and painful.

The second is that as disasters become less frequent, human nature is such as to push them entirely out of mind. That's the problem posed especially by volcanic eruptions, earthquakes, and tsunamis. By contrast, weather hazards such as cycles of flood and drought, riverine flooding, and hurricanes and tornadoes are more frequent and develop in plain view, and in the process they remain more in the public eye. Preparedness in the face of rare but high-consequence events is problematic. We're best at what we repeatedly experience and practice.

3. *Unintended consequences of our success will still be with us but will show up in new arenas.* Public health, the social safety net, and more will be affected. Just like science itself, the unintended consequences of technological advance and social change offer an endless frontier.

4. *Sustainability will remain a treadmill issue we can't get off; we'll still be innovating and using economic adjustments and substitutes to buy time for humanity.* Our efforts then will be more intentional and strategic than today's. We'll be better at the job of buying time, but we'll need to be. The challenge will seem greater. We might be using an alternative term: *Sisyphean* development, noting the similarity between the continuing need to buy time and the king of Greek mythology who was punished for his habitual deceit by being repeatedly forced to roll a huge boulder uphill, only to see it roll back down.

5. *Command-and-control leadership will have fallen out of favor.* The world's leaders will be more content to focus on broad policies and less inclined to micro-management and tight control, in part because the costs of these latter strategies will have been increasingly evident and persuasively documented and in part because IT and social networking will continue to drive devolution. (In this last respect, there's actually a data point. Consider this vignette: In the mid- to late 19th century, with the invention of the telegraph and submarine telegraphy, the British Foreign Office thought it would be able to assume greater home control on international dealings around the world. But they quickly learned that when it came to India, to cite just one example, the local representatives of British interests were able to manipulate home foreign office policies and decisions by controlling the flow of information from their respective posts.[4]) Nations will struggle to accomplish the transition from today's top-down, command-and-control world

---

4. For a fascinating narrative on this, see National Science Foundation's Vary Coates.

to the coming world where the power, means, and action are concentrated at and stem from local levels (where "local" here is taken to mean not just geographically but also with respect to professional disciplines, economic sectors, and more). The nations who best succeed at this task will likely be best poised to enter the 22nd century. (This reality poses particularly serious challenges for China, which has made considerable investments in regulating its citizens' access to and use of the Internet, through such initiatives as the Great Firewall, to restrict outside Internet access, and the Golden Shield, to control domestic content. Despite such efforts, China looks to lose the battle long-term.)

So this reanalysis, this closer look at the conditions and trends prevailing now, suggests that the perspective of 2100 is likely to differ substantially from that of today. The years 2000–2100 make up the pivotal century for the challenge of living on the real world.

12.2. How, Then, Will We Do Business?

Chapters 4–11 posit that to improve the 21st-century prospects, given our limited individual and corporate intelligence, constraints on our financial resources, and the relentless press of real-world challenges, humanity should focus on developing and mastering four tools: a basis of facts; policy; IT and social networks; and local-, place-based leadership. The assertion is that these tools are currently unequal to the task but that they are inexpensive individually and, in aggregate, amenable to or contribute to trial-and-error testing and improvement. They are inherently scalable and viral. They can quickly be upgraded and harnessed within the limited time available. What is the outlook for the four tools?

1. *A basis of facts.* Investment in Earth observations, science, and services (Earth OSS) will be substantially greater. This critical 21st-century infrastructure currently costs the world something like $30 billion annually (something like $10–$15 billion per year in the United States),[5] as compared with our current $70 trillion global GDP. The difference between what we see the world doing and what we know and understand about its behavior is like the difference between seeing and anticipating what hazards and opportunities lie ahead and flying blind into the 21st century. Currently we know enough to see some of what's in store, but only vaguely. The high stakes

---

5. These figures are vague but at least approximate. They merit more accurate calculation but suffice as estimates to show their small order of magnitude relative to the larger sweep of human affairs.

and urgency combine with the low costs to suggest that we try doubling our investment in this critical infrastructure over the next few years (taking that annual investment from 0.05 percent of global GDP to 0.1 percent of GDP) and see whether we like the results versus attempting to do some prior cost-benefit analysis or engage in partisan debate about the merits. Both consume time that we no longer have. It wouldn't be unrealistic to imagine a doubling every 5 years for the next 20 years, and then a slowing to a doubling every 10 years for the remainder of the 21st century. But even with that rate of growth, it will remain the same small fraction of GDP, no more than 1 percent to –2 percent, in part because the benefits from such an investment and accompanying acceleration of GDP growth will more than repay the costs. The benefits will be manifest. We'll begin to see an end to poverty and a world that increasingly has enough food, water, and energy to meet basic needs, for all peoples, everywhere. As current shortages disappear, increasing resources will be available for education and innovation. And as the contributions of the Earth sciences and related social sciences grow, public appreciation for their value will increase in like measure. We'll find ourselves on the virtuous side of Chapter 9's diagram.

2. *Policy*. Knowing what the earth has in store for us is not enough. We have to recognize our options. What can we do in response? And for each course of action, what will be the consequences? Will we have bought time or merely added to a growing bow wave of unaddressed problems? We acknowledge that we're not sufficiently intelligent individually or in small groups to tackle each challenge de novo. We need a policy framework to speed our decisions, to allow us to make decisions and take actions locally in ways that will navigate the short-term while at the same time aggregating to larger trends that will better position us for the future. Again, our current policy framework is not up to the task. So, in addition to specific details, the governing policy framework has to emphasize innovation and learning from experience and, toward that end, devolution of decision-making and action.

3. *IT and social networking*. The latter notion of devolution is not just an idle political preference. It is critical to 21st-century success. To increase dramatically the pace at which we can test ideas and learn from experience, we need to turn large numbers of local contexts into pilot projects, opportunities for learning quickly from experience and sharing that knowledge rapidly across society. In principle, IT and social networks offer precisely such possibilities.

4. *Leadership development*. The devolution of decision-making, the integration of research and practice in an infinitude of contexts worldwide, and

the exchange of information across these microcosms will transform leadership from its present form, better suited to the more hierarchical societies of the past, to leadership that is much more broadly distributed across society. The new leadership will be less about command-and-control principles and more about vision, integrity, and responsibility. It will be more deeply imbedded in problem-solving contexts on the ground, and show the dual characteristics of invitation and challenge. This local leadership will have a bias toward *action*, not mere contemplation. The good news is that large numbers of people are eager to shoulder the responsibilities that come with this role, but at the same time the numbers of those future leaders, at least initially, are far smaller than the world's seven billion. It's therefore realistic to envision inspiring, equipping, positioning, and supporting them in relatively short order.

Recall that each of these four approaches to coping with future challenges—a basis of facts; policy; IT and social networking; and local-, place-based leadership—is already underway. We're not talking about the need to create societal functions that are entirely new, or that require some sort of worldwide about-face. We're talking instead about aiding and abetting ongoing trends. We're talking about being intentional and strategic, not semiconsciously instinctive with regard to these matters. The needed policies should therefore be more readily put in play.

Overall, chances are good that in future years we will be living in the moment, thinking inside the box, pretty much as we are in reality right now. The difference will lie in our increased self-awareness and the speed and effectiveness with which we can bring such mindsets to bear on pressing problems. The Internet is only just beginning its transformation of society, how you and I live and work and relate to one another, and how we solve problems and move the world forward. Some national leaders from some countries and some corporations may cling to top-down, command-and-control approaches, but they and their nations and institutions will pay a price. They'll be left in the dust by a world that is embracing change and innovation and new approaches and solutions to every problem.

Some national leaders from some countries and some corporations may cling to top-down, command-and-control approaches, but they and their nations and institutions will pay a price.

That top-down approach will be replaced by a devolution that empowers individuals and small groups to work locally on bits and pieces of problems and opportunities within their respective boxes. We know little about what this will look like any more than the DARPA creators of the Internet could foresee the rise of Amazon, Google, Microsoft, Apple, Facebook, Twitter, Wikipedia, or crowdsourcing. Current crowdsourcing approaches such as crowd-voting, crowd-funding, micro-work, and creative crowdsourcing are just the earliest explorations of approaches that will be commonplace just decades from now.[6] The next big thing for IT is finding ways and means to harness new developments such as these to solving problems and making those problem-solving capabilities available to those "local" groups (the people thinking inside their respective boxes) that will want to draw on such resources. Expect, perhaps as early as a few years from now, a new set of household words—Web-based firms for crowdsourcing and for building tools for use by people and institutions at a local level—will be as iconic by 2100 as the Fortune 50 firms are today.

As for the work accomplished in the boxes themselves, it'll be quite different from issue to issue (water-resource management versus energy development and use versus the world of apparel or entertainment). The strengths and weaknesses of living for the moment and thinking inside the box will have been studied in detail and richly documented. (We'll have moved a couple of generations past those ideas to wholly new approaches and nomenclature, so the discussion will appear almost entirely unrelated to what's been presented here.) We'll have plenty of advice on how to make ourselves more capable. The establishment of local [place-based, topically focused] groups will be less haphazard, less accidental, and more premeditated. The rhythm of living in the moment and thinking within the box, punctuated by times of looking out more long-range and sharing information across boxes, will have become an accepted part of life and work. Communication and information sharing within and across boxes will be more highly regarded and understood. Innovation and learning from experience, not just person

---

6. An analogy to our current vantage point would be the first decade of aviation following the Wright brothers' historic flight. Planes were still made from wood and cloth. Military aviators were dropping explosives from planes by hand. Barnstormers were giving grandma a quick spin and a few loops above the farm and saying in the same breath "this is the people-mover of the future." Designers were deciding whether seven wings or twelve were optimal. Railroads looked at this activity and sneered. They decided their customers were more interested in whether or not the Chicago Zephyr and other trains of the time offered linen table service in their dining cars. That's why we see no airlines named Burlington Northern or Penn Central.

by person but focusing more deliberately on learning from the experience of others, will be integral to the process, and respect for these values will permeate educational systems and policy formulation worldwide. People will more broadly and effectively share what they know. (Note that this particular prediction depends on the development from some source of ways or means for providing incentives for such behavior, something analogous to and performing the same social function as patents, copyrights, and royalties but likely looking quite different.)

The importance of diversity of every kind—cultural, geographic, ethnic, gender, age, and more—will be universally appreciated rather than grudgingly conceded. We will master diversity and its benefits versus eliminate it through some great homogenization. In fact, that latter possibility will be recognized as a potentially risky outcome, one to be fended off. The essential importance to diversity in learning from experience and problem-solving will be recognized. We won't be able to afford seven billion people who share the same blind spot, and so we will try to protect and nurture diversity of human thought much as we now protect genetic diversity in food grains, trees, and more.

Most importantly, as we come to see that policy formulation works best when it's congruent with physical and social *realities*, not just idle wishes or fantasies, and as policies take on a more reality-based tinge, the policy discussion itself, at least with regard to living on the real world, will become less polarized and divisive, at every level of government and society.

Contention, however, will still be with us. It will simply have moved on to other issues. (My candidate? The allocation of extraordinary but necessarily limited and precious biotechnological and healthcare technology: Who has access to these? Only the wealthy? The best and brightest? Those chosen by lottery? Who gets to live? For how long? With what quality of life?)

And here's a proposition that might seem radical today (but my forecast is that it will seem tame in retrospect): The close relation of issues such as this to our most fundamental values suggests that the spiritual dimension of our lives—the dimension we've tried to suppress and put in a box for the past several generations—will command more of our interest and attention. The world will experience revival. (More about this in section 12.6.)

12.3. Downscaling: The Forecast for You and Me

So much for a 40,000-foot view of the future. What will it look like on the ground, at the individual level—that is, for people like you and me? Meteorologists work on a similar problem. They now run global climate models out hundreds of model years. These models capture variability on seasonal and interannual time scales, on decades, and on out to century-long trends. But in the present world, what matters to nations, to states, to corporations, and to other institutions is what happens on these time horizons at regional or local scales: northern Europe, say, or the Sahel, or Indonesia and Australia. In response, meteorologists have begun to develop a capability they call *downscaling*: an ability to synthesize the global results with additional detail on the particulars of local topography and other factors to translate that global picture into its implications for a region.

So how will life look to the individuals who populate this generation and this century? What will these larger trends portend for our personal lives and in the workplace? At the most positive end, increased awareness that there's a systematic approach in place and working to meet the resource needs of the world's people, protect the environment and ecosystems, and build resilience to present and future hazards holds the potential to enhance the meaning of everyone's life. Such awareness could also reduce the stress in our individual lives in two important ways. As we grow more confident that our larger collective policies are effective, we can shed today's pervasive vague sense of foreboding and the enervating pessimism that saps our enthusiasm and aspirations. At the same time, by simplifying the thought process we find in the workplace, by replacing the need for making resource, environmental, and hazards decisions on an individualized, ad hoc basis, we'll actually speed the process for deciding and taking action and improve the effectiveness of those decisions and actions at the same time. In these ways we'll remove a chronic source of stress that nags at all of us even in our quietest hours. This reduction in stress and more hopeful yet realistic outlook for the future will carry over into less frenetic, better-ordered lives that allow for and feature an improved life–work balance.

Part of the message for you and me is that as individuals, we can act on our own initiative to bring about this improved state of affairs. We don't have to wait for anyone else's permission. We don't have to change cities or jobs. We don't have to watch for the stars to align. We can start now. All we have to do is look at the four tools described here and give them a nudge, either singly or in combination:

1. A basis of facts about the natural and social world;
2. Policies based on these facts;

3. IT and social networking; and
4. Locally based, widely dispersed, responsible leadership.

We can gather a little more data on the natural or the social worlds. We can advance our understanding of how either or both of these worlds work. We can ponder the regulatory and policy framework that governs our jobs, our companies, our communities, our states, our countries. We can think through alternative possibilities. If we can provide some early assessment, perhaps we get "extra credit," but remember, as humans we're not that skilled in thinking through the emergent consequences, good or bad, intended or unintended, of policies. We're mostly going to be working our way out of our current predicament by trial and error. Just think of ideas that are new and/or that have emerged as lessons learned from experience.

But let's not stop there. Let's share these ideas. For example, if you can write a book or a paper, or a blog, that's great, but this is the era of social networking. You don't have to go that far. All you have to do is share with those closest to you. Your family. Your friends and neighbors. Your work colleagues. The people you *know* will listen, not the people you have to somehow *convince* to listen. That can come later, after those close to you have heard you out. Others are close to them vis-à-vis word of mouth. Especially if *you* make it a practice to listen to others around you. (Remember, *really* listen, as mentioned in Chapter 11. Don't make it a practice to listen only to the point where you can break in with something to say yourself.) And then share what you've heard. In the swarm intelligence of the 21st-century world, ideas don't have to be original with you. You simply have to pay them forward. And, finally, you don't have to replace your old social networking (letting your friends know where you are and what you're doing socially); you just have to add this slightly more substantive dimension to what you're already putting out there.

Again, don't stop there. With the ideas for which you feel a little passion, or see a special opportunity, push to bring them a little closer to reality.

Lastly, don't overdo. Your job is less to convince others than to be the change you want to see in the world. (This of course makes the assumption that the change you want to see in the world is not more overstressed, overworked, uptight people. They should see you as relaxed, as continually offering new ideas and standing behind them, but not insistent on having your own way. Believe me, their lives are already full enough with others who are plenty insistent.) You may not see your ideas bear fruit right away. You may never see your ideas in recognizable form. But you can be sure that your ideas and actions have influenced those around you and changed events.

12.4. The Outlook for Initiatives #1–#10

By and large, this book has been descriptive, suggesting a certain way of looking at events and trends on the real world. But a small fraction of the book has been devoted to a handful of suggestions. On the surface they might have seemed disparate and unconnected, but they share certain virtues. Like the four larger tools—a basis of facts, an evolving, ever-more-capable policy framework, rapid learning from social networking and independent pilot projects, and a proliferation of servant leaders on the ground—they're inexpensive. It doesn't cost much to give any one of them, or even all of them, a try. They support the four larger efforts. They'll do little harm if they are tried and don't work. The upside potential is far larger than the downside risk. They don't require the approval of a large segment of society before implementation. Individual institutions, or individuals for that matter, might give them a go. And if and when someone does, whether by the names given here or in some other manner recognizably connected, or in ways that appear quite independent of and unrelated to these suggestions, they have the potential to make or contribute to big societal changes, and to do so quickly, in the time needed (the first quarter or half of this century, say). They hold great potential to render *visible* this broad and uplifting challenge of living on the real world and making *tangible* the opportunities for millions of people to pitch in and contribute.

What about the outlook for them? How likely is it that there will be societal uptake?

First, it should be apparent from the entire tenor of the book that many of these notions are not so much original as already preexistent, at least in embryo . . . so it is more likely than not that most of them will be adopted and evident somewhere in society in some form within a few years. But some more specifics.

Real-World Living Initiative #1. In addition to speaking of sustainable development, let's initiate a number of analyses with respect to the needed natural resources that estimate how much time we have. Let's set up a limits clock, analogous to that Bulletin of Atomic Sciences doomsday clock, for each resource. Let's update those clocks periodically, celebrate those occasions when we buy ourselves some time, and focus our attention on those areas where time is short. (Chapter 2)

The *Bulletin of the Atomic Scientists* has already made a start here with respect to nuclear war, climate change, and biosecurity. Sustainable development could stand a new point of departure for analysis and thought. It would be surprising if something like this doesn't catch on.

Real-World Living Initiative #2. Double the rate of investment in Earth OSS in the United States (and hopefully, for that matter, worldwide). Reap the domestic benefits. But don't stop there. Develop and implement a Marshall Plan for the 21st century. (Chapter 5)

It's likely that the United States and other major nations will up their investments in Earth OSS and up them substantially. Whether they'll go the next step and make such critical infrastructure freely available to all other nations in the hopes of building a more sustainable world is more problematic. (An alternative scenario is that the investments are made but the information is privatized and used to further the self-interest of those who can afford to pay. It's not the post–World War II world. This would be a dangerous outcome.)

Real-World Living Initiative #3. The United States should build the capacity of CRS, CBO, and GAO and at the same time reestablish OTA. There should be some high-level consideration of ways and means to fund academic research in policy analysis in a more structured, formal, robust way, and in the process educate and train a generation of analysts who could employ those skills at state and local levels where they are badly needed, and who could provide the needed diversity and of disciplined thought with respect to policy formulation. (Chapter 7)

Something will happen here. But it may be confined to incremental budget plus-ups within the U.S. Congress (and possibly other world governments) for infrastructure such as the Congressional Budget Office, the Government Accountability Office, and the Congressional Research Service and possibly within the Social, Behavioral, and Economic Research division of the National Science Foundation (NSF). It's hard to see Congress making the big step from its current bitterly bipartisan stance to something more bold.

Real-World Living Initiative #4. Institute a basket of hazards policies (1) stressing no-adverse impact; (2) learning from experience and forming an independent National Disaster Reduction Board toward that end; (3) keeping score; (4) building public–private partnerships; (5) recognizing and reframing part of the Department of Commerce portfolio to achieve these goals; and (6) initiating pilot programs to identify and forestall global threats. (Chapter 8)

The growth in losses to natural hazards decade by decade is not only unsustainable but out of step with progress in other areas such as aviation and industrial safety. No-adverse-impact analogs appear in other areas of policy and in medicine. (For example, environmental impact statements are based on this premise. Medical doctors all take some form of the Hippocratic Oath, which includes the idea of do no harm.) Public–private partnerships remain a policy challenge but represent too important an opportunity to miss. Expect (finally!) to see some initiatives that work.

Real-World Living Initiative #5. Initiate a so-called L. F. Richardson World Outlook Project *worldwide, with the deliberate goal of modeling world outcomes (and downscaling to nations and regions) and their sensitivity to policy options with respect to resource use, environmental protection, hazard resilience, and more. (Chapter 9)*

The world is self-organizing along such lines at such a rapid rate that the reality of such projects is almost inevitable, and yet the variety and diversity of the initiatives will be so great as to make it hardly likely that a single name and individual would encompass them. Much as I'd like to see the name and reputation of L. F. Richardson elevated (for he embodies much to like in the frame of real-world thinking provided here), it doesn't look probable.

Real-World Living Initiative #6. Imbed a research/learning element into many of the current de facto policy innovation efforts already underway in governments, corporations, and communities of practice at every level worldwide and accelerate the rate at which knowledge on the relations among science, policy, and outcomes is developed and shared. (Chapter 10)

It's less likely that this will take the form of individuals designated as researchers and scholars imbedded in such work. It's more likely that the discipline of innovation will be more deeply instilled in every profession and community of practice. It will be accepted as a given that practitioners innovate and experiment, and evaluate their work in a disciplined, structured way.

Real-World Living Initiative #7. Develop opportunities for U.S. early career scientists to work abroad at the science-policy interface. This could be done by some combination of the U.S. Department of State and other federal agencies, much like the Fulbright U.S. Student Program, or by private funds, such as Fulbright Scholars. (Chapter 10)

This is too tempting an opportunity. One of the world's growing number of billionaires will almost certainly snap this up and claim it as his or her own legacy.

Real-World Living Initiative #8. Extend the NOAA/NWS Weather-Ready Nation initiative from an internal agency organizing principle to a United States that is truly weather ready at the community level. (Chapter 10)

This National Weather Service (NWS) initiative is rapidly finding traction and gaining acceptance. To turn this from a slogan or a federal-agency plan into a true description of U.S. resilience as a nation of communities to hazards seems a logical step. And the actions required—devolution of responsibility for planning and action to the community level; observations, science, and services focused at that level; supportive policy frameworks; and the need for nurturing of local leadership—seem natural in the context of the 21st century. A no-brainer and an obvious starting point for broader living on the real world objectives. Powerful incentives exist at the local level for implementing the needed actions. What's more, the program, though domestic, could be replicated internationally. In hindsight, we're likely to realize that this program became something of a portal or pilot project through which millions of people entered the larger work of living on the real world.

Real-World Living Initiative #9. Global dialog on three questions. 1. What kind of world is likely if we take no action? 2. What kind of world do we want? 3. What kind of world is possible if we act effectively? (Chapter 10)

The reality is that a worldwide dialog on these critical subjects is constantly in progress . . . but not under these simple headings. It might well be too ambitious (on the one hand) or too confining (on the other) to attempt to add such a conversation to those underway already. Yet, as argued in Chapter 10, such an attempt at a global synthesis might well perform a useful function, helping seven billion of us better see where matters stand.

Real-World Living Initiative #10. A new Smithsonian Museum. (Chapter 10)

Every museum in every land already offers exhibits and special showings endeavoring to promote a better public understanding of the interplay of resources, environment, and hazards. The remainder of the century will see a continuing stream of such offerings, of progressively greater scope and reach. The suspense lies in whether the United States (and its premier museum) will continue to promote and shape the national and worldwide dialog on these critical (indeed *defining*) matters for the 21st century, or whether it will leave such leadership to others.

12.5. Completing the Forecast—Uncertainty

To say that every subject and issue in this book is rife with uncertainty would be to understate. To say that I've failed to deal with any of those uncertainties would also be fair. I've tended instead to hide behind the Charles Darwin quote[7] and remind readers at every step that what's here is conjecture and viewpoint as opposed to established fact.

My fellow meteorologists have been raised to do better! In 2006, the National Academies of Science/National Research Council published a report titled "Completing the forecast: Characterizing and communicating uncertainty for better decisions using weather and climate forecasts." The report examined the state of communicating uncertainty by meteorologists and suggested ways the public and private sector might do more to accommodate this. The thrust of the report was that uncertainty is a fundamental part of every meteorological forecast and a critical element in how a given forecast can or should be used.

So, what about the uncertainty in the 21st-century forecast provided here?

Throughout this book, the message has been that seven billion of us find ourselves engaged in perhaps the greatest adventure of human history, one in which there are no bystanders, only participants, and one in which each of us is discovering along the way whether we're the heroes or the villains of our life's story (though we can and do tend to be *both*, depending on the different times or seasons in our lives).

It's an accidental adventure. We didn't go looking for it; it found us. For all of human experience we've faced *two* simultaneous problems: capturing resources, especially food, water, and energy, and surviving natural threats. After millennia, we finally found ourselves so adept at these two challenges that we inadvertently spawned a *third*: protecting the surface of the planet and its ability to sustain life from the harmful effects of seven billion of us knocking about. This newer, threefold problem is beyond our conscious comprehension. We certainly can't fully grasp it or think our way through it as individuals, and so far it has defied our corporate efforts to cope.

To this point, there would seem to be little room to quibble.

The uncertainty lies at the next step of the argument, that those four tools, if developed and harnessed appropriately, offer a way through. It's easy to see holes in this argument. Gathering data about the state of the natural and social world with the global coverage and level of detail needed could prove prohibitively expensive. Converting that information into knowledge and understanding might well take more time than nature will give us. Perhaps

---

7. "False facts are highly injurious to the progress of science, for they often endure long; but false views, if supported by some evidence, do little harm, for everyone takes a salutary pleasure in proving their falseness." (*The Origins of Man*, Chapter 6)

our new understanding will always be accomplished too late, only after the fact.

Today's policies clearly aren't up to the task. The proposal here, using IT and social networks to mount massively parallel policy development and advance at local levels on a global scale, is a long shot. National leaders will be reluctant to see power devolve. For its part, the public might see too much risk of instability. Even in the unlikely event that the schemes proposed here would be widely embraced and pursued with full vigor, they could well fall short. And it might be that humanity lacks leaders with the required talents and in the numbers needed to populate such an effort.

The arguments posed here are vulnerable to these criticisms and many more. All are serious. Any one failing will be enough to sink the effort.

Three rejoinders. First, there are few alternative approaches being offered to solve our threefold problem. Absent some better options, all we have to fall back on is improved documentation of human failure of ever-widening scope, increasing speed, and gathering complexity. Second, Adam Smith's unseen hand, the swarm intelligence, the wisdom of the crowd (you pick your favorite name), seems to be exploring the approaches outlined here. Every time you pick up a newspaper or magazine, click on any website, or surf any cable channels, you should be seeing signs of new startups, novel initiatives, and/or innovative policy proposals, with fresh, charismatic faces at the lead, addressing these problems in ways partially or wholly consistent with—or far more imaginative and far-reaching—than the ideas offered here.

And third, all the doubts and concerns we've discussed so far, substantial though they may be, are dwarfed by a final uncertainty, one that's both personal and generational.

12.6. The Uncertainty That Matters Most

Recall that living on the real world, especially the 21st-century real world, is an adventure.

The essence of adventure is uncertainty—and stakes and risk. Compare real-world war with World of Warcraft, the popular, massively multiple online role-playing game (MMORPG). It's the difference between adventure and cheap thrill-seeking. For that matter, think of the moral and ethical debate swirling around the use of drone aircraft in war. At the one end, real people are living and dying. At the other, real people are pounding a computer keyboard and moving a mouse and taking a break every now and then for a latte or to hit the gym.

In every such adventure, whether real world or virtual, the biggest challenge is never the enemy army or the dragon; or the terrorists or the mountain; or, as in the present instance, the threefold challenge of managing Earth's resources, environment, and natural hazards. Instead, the biggest challenge is always a test of the individual's character and spirit. And that is similarly the biggest uncertainty facing us over the next 100 years. How will you and I respond to this defining 21st-century challenge of living on the real world?

Jesus, when he was on Earth, delivered a message of universal love, eternal life, forgiveness of sins, and a resultant hope. Along the way he invited us to see and accept what he had on offer. He saw a range of options for how people respond to such big messages and challenges. He taught this parable (probably familiar to all of us, regardless of our faith).[8] He offered four categories:

> Listen! A farmer went out to sow his seed. As he was scattering the seed, some fell along the path, and the birds came and ate it up. Some fell on rocky places, where it did not have much soil. It sprang up quickly, because the soil was shallow. But when the sun came up, the plants were scorched, and they withered because they had no root. Other seed fell among thorns, which grew up and choked the plants, so that they did not bear grain. Still other seed fell on good soil. It came up, grew and produced a crop, some multiplying thirty, some sixty, some a hundred times.
>
> Then Jesus said, 'Whoever has ears to hear, let them hear.'

Later, in private, Jesus explained the parable to his disciples this way:

> The farmer sows the word. Some people are like seed along the path, where the word is sown. As soon as they hear it, Satan comes and takes away the word that was sown in them. Others, like seed sown on rocky places, hear the word and at once receive it with joy. But since they have no root, they last only a short time. When trouble or persecution comes because of the word, they quickly fall away. Still others, like seed sown among thorns, hear the word; but the worries of this life, the deceitfulness of wealth and the desires for other things come in and choke the word, making it unfruitful. Others, like seed sown on good soil, hear the word, accept it, and produce a crop—some thirty, some sixty, some a hundred times what was sown.

---

8. Mark 4, NIV.

To create the parallel, seven billion of us think about some aspect of the issues in this book—the world's resources, the environment, and natural hazards—each and every day.

But some are too poor and desperate to think about any more than food, shelter, and clothing for their family for the next few hours. A second group simply finds the issues too troubling; such thoughts make them nervous. They simply put the issues out of their mind. Still others (most of the world's better-off might probably self-identify with this third category) are preoccupied with living in and experiencing the pressures of today's virtual world, twice removed from these issues as explained in Chapter 3. They're distracted and/or hopelessly entangled in the day-to-day urgencies of 21st-century living. But there's a fourth group, challenged and impassioned by the underlying real-world problems and opportunities.

And they (you, me?) will accomplish great things.

The moment challenges entire generations to answer this call to character and spirit. But history and the realities of chaotic systems also indicate that for this to happen, it may be sufficient that any single individual or group of individuals comes to embody the needed change—that his or her (your or my) decisions and actions may provide the needed tipping point that carries the rest of us along.

The biggest uncertainty is therefore what *you're* going to do with this book. You could merely put it on the shelf or, if you're reading an e-book version, simply close the file. You could go on to reading or doing something else. But if you entirely ignore the tenets of this book, then mankind's prospects are diminished in corresponding measure. At the other extreme, you could treat this as a kind of workbook: ponder this or that passage; underline this or that particular bit; see where or how you plug in to one of the pieces of the puzzle laid out here—the resource extraction, environmental protection, hazards mitigation, policy development, emergency management, or education—and get at it. Alternatively, you could find the book frustrating or wrongheaded and use the aggravation to energize and motivate your superior synthesis of mankind's circumstance and prospects. You could choose instead to follow and otherwise encourage others around you in their efforts to think globally and act locally and make the world a better place. If you choose any of these latter pathways, especially one about sharing with others what you find compelling, what you find confusing, and what you are sure is wrong-footed, then we all have that much more reason to be more sanguine about our joint future.

Here's the other piece. If you and I don't rise to the occasion, the needed innovation and accomplishment and leadership may or may not come from some other quarter. Although the suspense may be less about whether history is moving in a certain direction and more about our role in it, do not allow yourself to feel complacent. Will future generations see you and me as agents of change? Or as dinosaurs that events passed by? Or somewhere in the middle?

12.7. Thinking Like a Meteorologist

Our reanalysis, our set of quick revisits to the logic of the book would not be complete without a look at what meteorologists have brought to the table—how thinking and acting like meteorologists have dramatically but not incrementally improved the human prospect. Meteorologists have taught us, by word and by example (modeling behavior):

1. How to see *what's coming for the rest of the 21st century*. Nothing improves our prospects like knowing what lies ahead. That's true at the individual level and corporately; it's true with respect to the stock market and every aspect of life; it's true for life as a whole. The meteorologist's persistence forecast, a simple tool, helps us see the 21st-century challenges for living on the real world with clarity.

2. *What's coming for the rest of the 21st century.* When we apply the idea of a persistence forecast, we discover that we can anticipate adverse climate variability and change; harsh extremes; resource scarcity and declining margins; myriad acute, localized crises arising from the unintended consequences of past action; and ineffective top-down decision-making and command-and-control action.

3. *We can't ignore any dimension, and we need to solve all dimensions of our problem* simultaneously. Meteorologists know when it's vital to be exact and when it's okay to be merely approximate. In weather prediction, it's necessary to consider *all* the atmospheric variables and solve for them all simultaneously. But it suffices to be approximate when it comes to stepping through how those simultaneous solutions will evolve place by place and step by step. To date, the world has tried to do something like the opposite, in two respects. First, we frequently choose to ignore one or more of the physical, social, and spiritual dimensions to reality and the way these three aspects play into virtually every important issue. Second, we've settled for tackling each of the three dimensions (resources, environment, hazards) to living on the real world in isolation: then, in that same isolated/incomplete context,

we've proffered unnecessarily prescriptive policy approaches that only look attractive because they ignore the other dimensions to the fuller problem. As a result, the so-called solutions for resources, environment, and hazard resilience turn out not to be solutions at all. And so-called solutions for living on the real world that fail to integrate the physical, social, and spiritual dimensions of our lives could stand a little improvement as well.

4. *We need not be afraid of/back off from/give up on insoluble problems.* The reason we've made the real-world living choices we've made so far—to treat resources, environment, and hazards in isolation—is that throughout our experience we've only had to deal with the two-part problem, and the solution has always been to go for the low-hanging fruit. Moreover, we've tended more to go it alone, not entirely alone but in small groups in competition with other groups. We've tended to cooperate when it's come to research and compete when it's come to living, when what we need to do is compete when it comes to research and cooperate when it comes to living.

5. How *to solve insoluble problems.* Meteorologists have shown us how to approximate and break up overall problems into local, momentary bits: *learn* from experience, *share* experience, innovate, and then try again and again.

6. *We should put every (new) tool to work and constantly improve the tools.* The history of meteorology from its origins has been more a recognition of how developments in other arenas might be applied and put to work on behalf of meteorology. This is true for observing instruments going back to thermometers, barometers, and more and extending to space-based radar, laser, and radiometric techniques. It's true for computation. Meteorologists have succeeded in part because they've seized on each new tool and platform for it that has come along and put it to work: thermometers, barometers, anemometers, hygrometers, radar, laser, radiometer, weather balloons, aircraft, buoys, satellites, and more. When meteorologists look at today's landscape they see the four tools of basis of facts (data from the observing instruments and research results, for example), policy (the computer/parallel processor), communication and social networking (dissemination), and leadership (the forecaster).

7. *The power of the individual working at the local level, even without perfect knowledge.* Lastly and most importantly, Lorenz's discovery of chaotic systems also helped meteorologists appreciate the power of the individual in a swarm. This perspective challenges all of humanity to see our importance while retaining our humanity and humility. It challenges all of us not just to stand for the change we want to see in the world but *to be* the change we want to see in the world.

EPILOGUE

All the world is a stage, and all the men and women merely players. —*William Shakespeare, from* As You Like It

Hard to disagree with the Bard's assessment, even four centuries later.[1]

Well, here's good news. For us, the stage is nothing less than the real world itself. And you and I are not bit players in some sitcom. Through God's grace or our own great luck, as opposed to any special merit, we've drawn major parts in a great drama. What's more, we've strolled on this world stage at a climactic moment, not an interlude. What we decide, what we say, and what we do—the big decisions and even the small choices we'll make the rest of our lives and especially how we choose to collaborate and work with each other—will matter and will shape human destiny for centuries to come.

We have landed on history's sweet spot.

For all of human experience until now, humanity has been along for the ride. The planet Earth we live on paid little attention to our ancestors. There was no need. Continuing with Shakespeare's metaphor, look back to Act I,

---

1. In the context of *As You Like It*, Shakespeare goes on to talk about the *Acts* or chapters in our individual lives. With apologies, here we're heading in a different direction.

the sweep of geologic processes, the formation of the oceans and the continents, the emergence of the present-day atmosphere, and the development of landscapes and watersheds. We weren't even on stage. Think of Act II, the development and evolution and spread of life—the complex molecules and the microbes—and the early life-forms. We still were nowhere to be seen, still no more than a gleam in the Creator's eye. By the time humans version 1.0 showed up, all of Earth's natural rhythms—the seasons of the year and the ocean currents and their year-to-year variations, and the ecological dynamics governing the plant and animal life—were pretty much established. Our human ancestors could do little more than capture some of the planet's benefits—the fresh water and the animals and the vegetation fit for food—and cower before the terrifying extremes—the floods and droughts and storms and cataclysmic earthquakes and volcanic eruptions. Humans version 1.0 had a *two*fold relationship: with the earth as resource and Earth as threat.

Fast-forward to today. Now there are seven billion of us, and our billions are chewing up resources at a per capita rate 10 to 100 times greater than did those who came before. (It's as if the planet were populated by 70 to 100 billion of the old humans version 1.0.)

In the blink of an eye, in the span of one or two human lifetimes, we have gone from inconsequential to the biggest actors on the surface of Earth. We're told that to the geological *periods*, the Cambrian, Ordovician, Jurassic, Cretaceous, and those more recent *epochs*, the Paleocene, Pliocene, Pleistocene, we should now add the *Anthropocene*: the age of mankind. Now humanity (humans version 2.0) matters, because we're seven billion strong. And we've added a third dimension to our relationship with the real world: protecting Earth as victim.

Although we mean no harm, we're discovering that nowadays when we feed ourselves, it's on such a grand scale that we take away from what had been available for other plants and animals—on land and in the sea. When we slake our thirst and when we use water to grow things and in industrial processes, we're drawing down freshwater supplies and we're rendering a lot of that water undrinkable. When we tap the earth's energy, we're polluting the air and the water and the ground. It's obvious that we can't keep this up.

No generation before us has faced this threefold challenge.

And the generations that follow will find that those of us living today have shaped and constrained all the options they'll have going forward. The decisions we're making each day and the actions we're taking now will determine whether the play we're all in heads toward a happy ending or a tragic one. Those future generations will either thank and revere us or regard us a little less favorably. The good news? They're unlikely to see us as sinister or evil. They won't see us as deliberately acting against their interests. But there's some chance they might see us as unthinking—or blind to the realities about us and to the implications for them—or too ignorant or lackadaisical or cowardly or selfish to measure up to what our times required.

If you're old enough, this might remind you of a wonderful book written in 1998 by the television broadcast news reporter Tom Brokaw, titled *The Greatest Generation*. He was referring to the cohort of Americans who grew up during the Great Depression, fought in World War II, and returned home to build America into a 20th-century superpower (starting, as you'll recall from Chapter 5, with generosity, by spending 2.5 percent of U.S. gross domestic product [GDP] each year over a period of four years to help the enemy they'd just defeated). Mr. Brokaw summarized the period not through some kind of bland overview but from the narratives of individual participants. Some were men and women we've never heard of, such as his parents, but others such as George McGovern, Robert Dole, and Daniel Inouye went on to become senators or people of equal stature.

Brokaw's label stuck. Ever since, people have known this demographic group as the Greatest Generation. As of this writing, those who are still alive are in their nineties; with each observance of Memorial Day or Veterans Day, we're told how their ranks have shrunk with the passage of time. Years later, we're still happy to acknowledge them as the best ever. But part of Mr. Brokaw's story is that they didn't see themselves as anything special. They simply "did what had to be done," working instead of playing as children to keep the family together during the Depression; risking and losing limbs, sight, and sometimes their very lives in remote lands they'd never heard of during the war years; and coming back to forge a new start in a country with little in the way of a social safety net.

But let me venture an appraisal. If our generation meets our food, water, and energy needs for the coming century in ways that protect our environment, habit, ecosystems, and biodiversity; if along the way we can learn to live with extremes in a manner that makes disasters a thing of the past; if, as seems possible,[2] we can reduce poverty to an irreducible 2 percent or less; if we can accomplish these goals not just for a single nation but for the entire world: We will have faced down a physical, social, and spiritual challenge

2. The June 1, 2013, issue of *The Economist* provides a highly readable cover story/special section on this topic.

that makes dealing with the Great Depression, fighting World War II, and making one country of many into a superpower seem trivial by comparison.

In this spirit, here's the book's last forecast, extending out 200 or 300 years. It starts with how people of that future moment will probably see us. They will look back on this period of history—centered around the year 2000—and realize that this age, no less than the Bronze Age or the Age of Exploration or the Age of Reason, had a name. I'm not smart enough to coin it, but it'll be something like the Age of Global Stewardship or the Age of the Environment or the Age of Earthkeeping, with a little more zip. They'll look back at us and say:

> Around the year 2000 our ancestors built instruments and sensors and satellite platforms, and they harnessed other technologies to observe Earth both as a whole and with respect to the smallest detail. The information technology they developed allowed them to capture and to digest the data. There was a flowering of understanding, and it came in the nick of time because seven billion people had been turning the planet into an ashcan. Because of what they learned and what they did, our Earth is livable today.

Those same future generations will envy us. (They might lift that mantle of the Greatest Generation off the shoulders of our ancestors and lay it on us. Or they might even give us a new name, call us the first humans version 3.0.) Further, they'll look back and say:

> It must have been great to be Jane or Rafael or Stanislav or Bamidele or Hui Zhong or Kwang-Sun or Dan or Dietrich or Lev or Giselle or _____ (insert your first name here). Not only were they living and working during this pivotal time, but they had this sense of manifest destiny. They knew how it would all turn out.

They'll go on to lament:

> It's not like today. Today all we have is declining budgets, organizational turf battles, personality squabbles and infighting, and a morass of empty red tape. Our work is meaningless, our play doesn't relieve our tension, and our worship brings little peace.

They'll forget what it felt like to be a member of Columbus's crew during the Age of Exploration. You'd been shanghaied on board, the work was back-breaking, the hold was dark, the bed was damp, the bread was moldy, the grog was weak, and the captain was certifiably crazy. And the guy on the bunk above you was disturbing your sleep, going on and on about some golden age of discovery.

Here's another small story, this one about the Greatest Generation, and how it was experienced on the ground. It comes from my uncle, the plasma physicist:

In 1943 he was about 13 years old, and living in Greensboro, North Carolina, one of the largest wartime concentrations of soldiers anywhere in the country. He was in a movie theater, mostly filled with soldiers, where they were showing a film titled *For the Duration*, referring to the duration of the war. The film had a subtitle: *1941–194?* When the subtitle appeared, as he tells it, the soldiers actually jeered. No one in the theater thought the war could possibly end before the end of the 1940s. He experienced a frisson, realizing that in such an extended war he'd be drawn into the combat himself.

But within two years, the war was over, and soldiers were coming home.

So do you have reservations about humanity's ability to meet its resource needs for today and at the same time protect the environment and ecosystems and guard against natural hazards? Have you been frustrated every day of your adult life by the slow and intermittent pace of any progress, contrasted with devastating, almost daily setbacks and defeats to your work on these fronts?

Here's the physical, social, and spiritual reality to the great, unprecedented, never-to-be-repeated adventure that you and I are sharing: The things that discourage us in this way, daily, year upon year will one day be regarded as our greatest achievements.

This just might be how it feels when things are going well.

Be strong and very courageous![3]

---

3. God's instruction to Joshua as he and the Israelites enter the Promised Land. (Joshua 1:7 NIV)

ACKNOWLEDGMENTS

> I'll confess at the outset that I wrote this book mainly in order to write an effusive set of acknowledgments. —*Juliet Eilperin*[1]

Ms. Eilperin got it right. However, when you're older, writing a book presents two special hazards: (1) over the years, you have accumulated a larger number of people to thank, and (2) you have had more time to reflect on, and truly appreciate, what a great debt you owe to each of them.

A balanced acknowledgment could therefore easily grow lengthier than this book itself. How to avoid starting down that road and yet at the same time convey my gratitude?

Some have suggested that I limit the thanks here solely to those people who directly helped bring the book to fruition. That would certainly work, but seems too restrictive, especially since the Author's Intent argues that essentially none of what we're prone to call "our" thoughts and ideas actually originate within us. Instead, most are grafted-on composites of thoughts and

---

1. In the wonderfully refreshing and spirited acknowledgments to her book *Fight Club Politics: How Partisanship is Poisoning the House of Representatives* (2006), Rowman and Littlefield.

views and knowledge we've acquired from others and subsequently reshaped and refined (or even misshaped and distorted).

The other extreme (mindful of Ed Lorenz's butterfly that each moment, by merely fluttering about, is constantly and forever changing the world's weather) is to thank every man, woman, and child of the seven billion people alive today, and every human being who has ever lived. Please reflect on this a moment. The chances are that you and I have never met. But chances are equally good that no matter what you're doing or where you're from, you've *transformed* a handful of people nearest you, who've in turn *influenced* their family, colleagues, and neighbors, who've then in turn worked through a surprisingly short chain of others to reach thousands and then millions, and in the process enriched my life and thinking. How can I thank *you*? A bit of a challenge, but perhaps we can go about it this way. Please enter your name here:

My thanks to _____, who, whether wittingly or unknowingly, influenced the thinking in this book, affected my life, changed me for the better, and/or will do so in the future, as this book or some other circumstance of life brings us together.

Quick, to be sure . . . but not entirely satisfying.

Some of you have known me by name and changed and blessed me by your words and example more than I can ever repay. And you know it! You know who you are and what you've meant to me. I need to recognize a few of you by name in return, as proxies for the much larger group I'd like to thank personally (and making apologies at the same time to those whose names are omitted).

That starts with family. We're told that our most important decision in life is our choice of parents, and I chose really, really well. My father taught me a little mathematics but more importantly modeled a quiet honesty and respect for others. He was brilliant, but the people who worked for him revered him as a human being, and growing up I got to see and admire that. My mother taught my younger brother and me how to understand my dad, and through that lens how to do life. She was so modest and unassuming in going about this that it's only belatedly, over the past decade or so, that I've come to appreciate fully what an enormous debt I owe her. My younger and ultimately smarter, richer, and better looking, but forever a quarter-inch shorter (no sibling rivalry in our family!) brother first showed me an alternative way of dealing with our parents and then how to carry oneself in triumph and trial. I've learned a lot from him and in retrospect should have learned more. Through these three I'm related by DNA and by marriage(s) to the most wonderful ensemble of family members and in-laws a man could ask for, starting with my wife Chris, my stepson Sean and his wife Shawn, my son Malcolm, my daughter Amanda and her husband Rob, and culminating in four grandsons: in birth order, Charlie, Jackson, Jamie, and Grayson.

Next is my formal education; here I owe a special thanks to Swarthmore College, its culture, its honors program, and my professors there and to Professor Colin O. Hines, my thesis advisor at the University of Chicago.

Three great institutions have kept me occupied and off the streets over the years. First in line was the Environmental Science Services Administration, now the National Oceanic and Atmospheric Administration of the U.S. Department of Commerce, where I worked in various positions in Boulder, Colorado, and Washington, D.C., for 32 years. There I learned one of the world's best-kept secrets—federal work is profoundly satisfying. Serving the American people? Using science to build public safety, grow the economy, protect the environment, and help the U.S. be a good neighbor among the nations of a changing world? Hard to find a better calling. The term "civil servant" has majesty and magic to it. Second, for seventeen of my twenty years with NOAA in Boulder I also served gratis as an adjoint faculty member in what used to be the Department of AstroGeophysics at the University of Colorado, teaching courses and supervising students; I'll forever be grateful to CU for this opportunity. Since 2000 I've been at the American Meteorological Society.

In these organizations I've had a series of extraordinary bosses, including, in more or less chronological order: (at the U.S. Department of Commerce) Lester Berry, Earl Gossard, Gordon Little, Vernon Derr, George Ludwig, Joseph Fletcher, Melvin Peterson, Ned Ostenso, Sylvia Earle, John Knauss, D. James Baker, Kathryn Sullivan, Alan Thomas, Chandrakant Bhumralkar, David Evans, (and at the American Meteorological Society) Richard Greenfield, Ron McPherson, and Keith Seitter. They each showed me, in different respects, but always by example, what it means to be a manager and a leader. They were also generous: when it came to annual appraisals, they evaluated me less on evidence-based measures of past performance and more in terms of roseate estimates of future potential.

Then come the hundreds of friends, fellow students and co-workers and collaborators from my jobs and from dozens of agencies, companies, and universities . . . what a contingent of bright, high minded, accomplished people! As a surrogate for this entire group, I'd like to zero in on the Policy

Program staff here at the AMS: Mona Behl, Judsen Bruzgul, Caitlin Buzzas, Tory Colvin, Peter Cowan, Tammy Dickinson, Joe DiTommaso, Gina Eosco, Genene Fisher, Paul Higgins, Ellen Klicka, George Leopold, Carolyn McMahon, Jinni Meehan, Shalini Mohleji, Anthony Socci, Jonah Steinbuck, Wendy Marie Thomas, Jan Wilkerson, and Tyna Wright. Some have moved on; some have joined us only recently. Some were kind enough to read earlier drafts of this book and offer comments. All have expanded my horizons and improved my thinking.

When it comes to expanding my horizons, I owe a special debt to a community of social scientists who have taken the time to give me remedial instruction in their respective disciplines. Their ranks include, in alphabetical order: William Anderson, Ian Burton, Heywood Fleisig, Shirley Laska, Jeffrey Lazo, Molly Macauley, Dennis Mileti, Robert O'Connor, Kristina Peterson (and her husband Richard Krajeski), Roger Pielke Jr., Kathleen Tierney, Dennis Wenger, and Gilbert White. Speaking of improving my thinking, no one has worked harder at this task for the past 13 years (and, truth be told, off and on for more than a decade prior) than Richard E. Hallgren. He's probably seen it as a Sisyphean labor; I owe him a unique debt. He's given me hundreds of hours of his time, and each hour has been a learning experience and a pleasure.

When I first arrived at the AMS in June of 2000, Ron McPherson, who then was executive director, asked me (and gave me broad latitude) to conceive, develop, and run an annual summer leadership program. The result was the ten-day annual AMS Summer Policy Colloquium. In the years since, over 400 early-career professionals and graduate students from the public, private, and academic sectors have taken part. Some 300 speakers and panelists, including members and staff of Congress, science advisors to the president, federal policy officials, journalists, international figures, and other science policy leaders have passed through to share their knowledge and experience and to inspire. Getting to know, work with, and learn from these remarkable people as their careers have developed has been one of the most satisfying experiences of my life. All seven hundred of us owe a special debt to Dr. David Verardo and the National Science Foundation, who provided funding that has allowed participation of graduate students and some university faculty (selected through a national competition) over this period and who provided helpful advice and several spellbinding lectures to the group as well. It's fair to say that without his support we'd have been forced to fold.

Along the way, the AMS Policy Program and its members have received additional support from NASA, NOAA, DoE, DHS, USDA, EPA, and other federal agencies and from private corporations such as Lockheed Martin, Ball Aerospace, ITT-Exelis, Northrup Grumman, Raytheon, and SAIC, as well as Google, the Lounsbery Foundation, the Rockefeller Brothers Foundation and from private donors.

My church community in Colorado and here in Virginia has supported me mentally, emotionally, and spiritually as well. I owe a special debt to a succession of pastors over the years: (in chronological order) Vernon Martin, Bob McPherson, Jim Andrews, Bob Westenbrook, John Terpstra, Andy De Jong, Mark Tidd, Don Bowen, Dale Seley, and Dan Carlton. Dan has been especially influential, by turns challenging and inviting me to weave together and profess (as opposed to compartmentalize and conceal) the twin threads of science and faith that make me who I am. And that's just the pastors. Space doesn't allow me to acknowledge hundreds of additional friends by name, but in the context of this book I need to thank one: Sara Bridwell, who in 2010 said, *Bill Hooke, you should start a blog!*

Living on the Real World (http://livingontherealworld.org) was the result. What a learning experience! And what an entry into today's world of social media! The comments from readers have taught me a lot; the occasional link from other bloggers has meant a great deal. What's more, though I'd been thinking idly of writing this book or something like it for decades, the discipline and process of developing the hundreds of posts for the blog over the past few years helped make the project concrete.

Which brings me to the thanks I owe for help with the book itself. Ken Heideman, AMS Director of Publications, and Sarah Jane Shangraw, AMS Books Managing Editor, at many pivotal points could have said "no" but said "yes." Together and individually they followed up, helping break down a mammoth undertaking into manageable parts, leading me to throw away horribly flawed false starts and get back on track while keeping me positive and motivated, and generally making the whole thing fun. The production editor Beth Dayton, the copy editor Jocelyn Humelsine, the proofreader Roger Wood, the cover and interior designer/compositor Eric Edstam, and indexer Nancy Peterson all did a masterful job of transforming the rough manuscript into something far more polished, while maintaining the intent.[2] Keith Seitter (I think of all my bosses, I may have saved the best for last) has been not just patient but supportive throughout. And don't forget those members of the Policy Program staff who kindly read and commented on the

---

2. Reminds me of the old recipe for making a silk purse out of a sow's ear: step one. Take a piece of silk.

earlier version. All these people have helped make the book better. (But not perfect! The book still has flaws—some I know about, and others remaining to be discovered. Those deficiencies are mine and mine alone.)

Finally, there's one person whose help and support eclipses that of all the others: my wife Chris. Writers, especially struggling ones, don't make the best company. And their scratchwork doesn't make the most scintillating reading. Chris has read and commented on much of the material, graciously putting aside much friendlier literature beckoning on her Kindle. She's amiably and patiently tolerated the sight of a distracted, oblivious husband, crouched over the computer evening after evening and weekend after weekend for months on end.

Thank you, honey.

INDEX

U.S. Army Signal Service, 65

U.S. Census Bureau, 58, 103, 110, 140

U.S. Clean Air Act, 65

U.S. Coast Guard, 192–193

U.S. Department of Agriculture, 71, 98, 99

U.S. Department of Commerce, 58, 111–112, 138, 140–141, 146

U.S. Department of Defense, 98

U.S. Department of Education, 99

U.S. Department of Energy, 71, 99

U.S. Department of Human Services, 99

U.S. Department of Interior, 71

U.S. Geological Survey, 71, 99

urbanization, 32, 91–92

Victorian Internet, xxv, 178

vision, 194–198

volcanic eruptions, 17

Washington Post, 211

waste disposal, 20–21

water, per capita consumption of, 44–45

watershed management, 67, 108

Watson, James, 47

Wave Propagation Laboratory (WPL), 196–197

weather balloons, 77

weather data, sharing of, 178–180

Weather Prediction by Numerical Process, 159–160

Weather-Ready Nation program, 184–185, 225

Weathering the Storm (Fleming), 212

What Went Wrong? (Lewis), 116

White, Gilbert, 73, 132

wicked problems, 36–38, 41, 88

Willis, Edmund, 65

Wilson, Edward O., 27, 86

Witt, James Lee, 139–140, 185

women, equal opportunity for, 138

Woods Hole Oceanographic Institution, 99

World Business Council for Sustainable Development (WBCSD), 174

World Meteorological Organization (WMO), 179, 180

World of Warcraft, 228

zero-margin world, 55–56, 129

Zuckerberg, Mark, 203